Mining, Mobility, and Social Change in the Global South

This volume focuses on how, why, under what conditions, and with what effects people move across space in relation to mining, asking how a focus on spatial mobility can aid scholars and policymakers in understanding the complex relation between mining and social change.

This collection centers the concept of mobility to address the diversity of mining-related population movements as well as the agency of people engaged in these movements. This volume opens by introducing both the historical context and conceptual tools for analyzing the mining-mobility nexus, followed by case study chapters focusing on three regions with significant histories of mineral extraction and where mining currently plays an important role in socio-economic life: the Andes, Central and West Africa, and Melanesia. Written by authors with expertise in diverse fields, including anthropology, development studies, geography, and history, case study chapters address areas of both large- and small-scale mining. They explore the historical-geographical factors shaping mining-related mobilities, the meanings people attach to these movements, and the relations between people's mobility practices and the flows of other things put in motion by mining, including capital, ideas, technologies, and toxic contamination. The result is an important volume that provides fresh insights into the social geographies and spatial politics of extraction.

This book will be of great interest to students and scholars of mining and the extractive industries, spatial politics and geography, mobility and migration, development, and the social and environmental dimensions of natural resources more generally.

Gerardo Castillo Guzmán is Associate Professor of Anthropology at Pontificia Universidad Católica del Perú (PUCP), Peru. He is Coordinator of the Anthropology of the City Research Group at PUCP and Honorary Research Fellow at the Centre for Social Responsibility in Mining at the University of Queensland's Sustainable Minerals Institute, Australia.

Matthew Himley is Professor of Geography at Illinois State University, USA. He is a nature–society geographer with research interests in the political ecology and political economy of resource industries, especially in the Andean region of South America. He is Co-editor of *The Routledge Handbook of Critical Resource Geography* (Routledge, 2021).

David Brereton is Emeritus Professor at the University of Queensland, Australia, where he was Foundation Director of the Centre for Social Responsibility in Mining. Since retiring from the University in 2016, he has continued to undertake research and advisory work focused on improving corporate social performance in the global mining sector.

Routledge Studies of the Extractive Industries and Sustainable Development

Andean States and the Resource Curse
Institutional Change in Extractive Economies
Edited by Gerardo Damonte and Bettina Schorr

Stakeholders, Sustainable Development Policies and the Coal Mining Industry
Perspectives from Europe and the Commonwealth of Independent States
Izabela Jonek-Kowalska, Radosław Wolniak, Oksana A. Marinina and Tatyana V. Ponomarenko

The Social Impacts of Mine Closure in South Africa
Housing Policy and Place Attachment
Lochner Marais

Local Communities and the Mining Industry
Economic Potential and Social and Environmental Responsibilities
Edited by Nicolas D. Brunet and Sheri Longboat

The Shaping of Greenland's Resource Spaces
Environment, Territory, Geo-Security
Mark Nuttall

Indigenous Responses to Mining in Post-Conflict Colombia
Violence, Repression and Peaceful Resistance
Diana Carolina Arbeláez Ruiz

Mining, Mobility, and Social Change in the Global South
Regional Perspectives
Edited by Gerardo Castillo Guzmán, Matthew Himley, and David Brereton

For more information about this series, please visit: www.routledge.com/Routledge-Studies-of-the-Extractive-Industries-and-Sustainable-Development/book-series/REISD

Mining, Mobility, and Social Change in the Global South

Regional Perspectives

Edited by
Gerardo Castillo Guzmán, Matthew Himley, and David Brereton

Routledge
Taylor & Francis Group
LONDON AND NEW YORK

earthscan
from Routledge

First published 2024
by Routledge
4 Park Square, Milton Park, Abingdon, Oxon OX14 4RN

and by Routledge
605 Third Avenue, New York, NY 10158

Routledge is an imprint of the Taylor & Francis Group, an informa business

British Library Cataloguing-in-Publication Data
A catalogue record for this book is available from the British Library

ISBN: 978-1-032-32179-0 (hbk)
ISBN: 978-1-032-32182-0 (pbk)
ISBN: 978-1-003-31323-6 (ebk)

DOI: 10.4324/9781003313236

The publication of the Open Access version of this book was funded by The Nordic Africa Institute

Contents

Illustrations

Figures

Tables

Notes on contributors

About the editors

David Brereton is Emeritus Professor at the University of Queensland, where he was Foundation Director of the Centre for Social Responsibility in Mining. Since retiring from the University in 2016, he has continued to undertake research and advisory work focused on improving corporate social performance in the global mining sector.

Gerardo Castillo Guzmán is Associate Professor of Anthropology at Pontificia Universidad Católica del Perú (PUCP). He is Coordinator of the Anthropology of the City Research Group at PUCP and Honorary Research Fellow at the Centre for Social Responsibility in Mining at the University of Queensland's Sustainable Minerals Institute.

Matthew Himley is Professor of Geography at Illinois State University. He is a nature–society geographer with research interests in the political ecology and political economy of resource industries, especially in the Andean region of South America. He is Co-editor of *The Routledge Handbook of Critical Resource Geography*.

About the authors

Professor Glenn Banks is a geographer who has researched and published on issues of extraction and development in Melanesia since the late 1980s, in addition to carrying out consultancy and policy work across the region for companies, communities, the state, and international institutions such as the United Nations Development Programme.

Séverine Bouard is a human geographer (PhD) at the New Caledonian Agronomic Institute. Her research focuses on assessing agricultural practices among Indigenous Pacific people in the context of nature's commodification and emerging indigenous discourses on nature and place. Focusing on the trajectories of people and territories, she highlights the social changes at work.

Valentine Boudjema is a PhD student in human geography and anthropology at the Institute of Research for Development and the University of New Caledonia. Her thesis examines the socio-spatialization of extractive and metallurgical activities and questions the notion of mining enclaves by highlighting the material and non-material circulations between mining and non-mining spaces.

Muriel Champy, Head of the Department of Anthropology at Aix-Marseille University and Member of the Institut des Mondes Africains, explores the economic strategies and imaginaries of success that are developed by youth and that are excluded from classical models of social ascension in West Africa.

Gerardo Damonte is Professor in the Department of Social Sciences at Pontificia Universidad Católica del Perú (PUCP). He holds a PhD in anthropology from Cornell University. Currently he is the director (PUCP) of trAndeS—Postgraduate Program on Social Inequalities and Sustainable Development in the Andean Region. His research addresses socio-environmental issues in the Andean region, particularly social dynamics linked to extractive developments.

Anna Dessertine is Research Fellow in social anthropology at the French National Institute for Sustainable Development (UMR 215 PRODIG) and is currently assigned to Rabat, Morocco at the National Institute of Planning and Urbanism. Her work aims to understand the socio-spatial mutations of artisanal and small-scale mining from a territorial perspective in Guinea, Ivory Coast, and Morocco.

Ibrahima Kalil Doumbouya is a PhD student, jointly supervised through the Higher Institute of Mining and Geology of Boke, in Guinea, and IMT Mines Alès (HydroScience Montpellier), in France. His work aims to analyze the socio-environmental and spatial dynamics of gold sites according to a multiscalar and multidisciplinary approach.

Philippe Dunia Kabunga is Lecturer at Institut Supérieur de Développement Rural in Kaziba and a researcher at the Centre d'Expertise en Gestion Minière at the Université Catholique de Bukavu, DRC. He is interested in participatory governance and the analysis of social and environmental conflicts around mining projects.

Sara Geenen is Associate Professor in globalization at the Institute of Development Policy, University of Antwerp, Belgium. She is Co-director of the Centre d'Expertise en Gestion Minière at the Université Catholique de Bukavu, DRC. Her current research interests lie in the global and local development dimensions of extractivist projects, addressing questions about more socially responsible and inclusive forms of globalization.

Julieta Godfrid obtained a PhD in social science studies from Universidad de Buenos Aires. Her research interests span sociology and international development.

Her current research focuses on corporate social responsibility, socio-environmental conflicts, and state-firm relations in Latin America, particularly in Argentina and Chile. She is a postdoctoral researcher at Universidad Autónoma de Chile.

Pierre-Yves Le Meur is an anthropologist and senior researcher at the French Research Institute for Sustainable Development (UMR SENS), based in New Caledonia (2008–2015 and since 2021). He works on the politics of belongings and natural resources (land, minerals, salt and freshwater) in New Caledonia and the South Pacific.

Ana Paula López holds a degree in geography and environment from Pontificia Universidad Católica del Perú and is pursuing a master's degree in water resources management at the same university. She is currently researching the relationship between extractive industries, socio-technical disputes, and environmental health in the Andean region of Peru.

Simon Marijsse holds a PhD in development studies and anthropology from the Institute of Development Policy, University of Antwerp, and the Department of Social and Cultural Anthropology, KU Leuven. His research lies at the intersection between the anthropology of mining and of technology.

Andrea Marston is Assistant Professor in the Department of Geography at Rutgers University. Her research explores the politics, labor, and technologies of resource extraction, with a focus on mining in Bolivia. Her book, *Subterranean Matters: Cooperative Mining and Resource Nationalism in Plurinational Bolivia*, is forthcoming with Duke University Press.

Patience Mususa is Senior Researcher at the Nordic Africa Institute, Uppsala. An anthropologist, focusing on southern central Africa, she studies mining and urbanization, urban climate politics, and Africa's development agenda for its battery minerals.

Iva Peša is Assistant Professor in Contemporary History at the University of Groningen, specializing in African environmental history. She leads a European Research Council-funded project "AFREXTRACT: Environmental Histories of Resource Extraction in Africa" (2022–2027). Iva's work draws from long-term oral history and archival research, particularly in Zambia.

Robin Petit-Roulet is a PhD student in geography at Institut de Recherches et d'Applications des Méthodes de développement (IRAM) and at the UMR Prodig (University of Paris 1 Panthéon-Sorbonne). His work focuses on the interactions, through mobility, between agricultural dynamics and gold mining development in Guinea.

Tobias Schwörer is a social anthropologist with more than three years of fieldwork experience in Papua New Guinea. He is currently Senior Lecturer at the University of Lucerne and has published on issues of extractive industries, resource conflicts, state-society interactions, and social inequalities.

Acknowledgments

To begin, we are extremely grateful to our contributing authors for their interest in the project, for the quality of their contributions, and for their cooperation and forbearance through the editing process. It has been a pleasure to work with you, and we have learned much from you.

As we crafted our proposal for this project and began recruiting contributors, we received valuable advice and encouragement from many colleagues. We especially wish to thank Nick Bainton, Tony Bebbington, Gerardo Damonte, Gavin Hilson, Leah Horowitz, Rita Kesselring, Miles Larmer, Andrea Marston, Ben Radley, Benjamin Rubbers, and Judith Verweijen. We are also grateful to Hannah Ferguson at Routledge for supporting the project and for sage counsel during the proposal-development stage, and to three blind referees for their valuable feedback on our submitted proposal, which improved the end product significantly.

Katie Stokes at Routledge provided advice, assistance, and encouragement as we moved through the writing and editorial process. We are grateful to Marilyn Ishikawa for making maps for three chapters and a figure for one, to Laura Soria for elaborating figures for one chapter, and to Priscila Arbulú, Kari O'Doran, and Rosemary Underhay for editorial assistance. The Vice-Rectorate for Research at Pontificia Universidad Católica del Perú provided funds for mapmaking and professional editorial assistance. Illinois State University provided funds for the index.

For their love, patience, and support, David thanks Erica, Gerardo thanks Laura, and Matt thanks Reecia, Jona, and Ellis.

1 An introduction to mining, mobility, and social change

Matthew Himley, David Brereton and Gerardo Castillo Guzmán

Introduction

Mining is movement. Pickaxes are swung, rock blasted, river bottoms dredged, ore trucked and processed. At a basic level, mining is the act of removing valuable materials from the earth, so these may be circulated through systems of economic exchange and re-embedded into myriad socio-technical assemblages: from cell phones to drywall, car batteries to food packaging. Movement is also critical to creating the *conditions* for mining to occur. Ore deposits form when mineral-forming elements and compounds are concentrated, through dynamic processes such as evaporation or crystallization, from their (mobile) host fluids (e.g., groundwater, seawater, magma). The search for commercially viable deposits drives the movement of people (e.g., drill operators, geologists) and things (e.g., maps, samples) across often-vast distances. To then extract minerals from the earth – and to separate their valuable components from their non-valuable ones – a wide variety of things are put into motion (i.e., "mobilized"): things like capital, chemicals, knowledge, labor, and technologies. Once the target element has been isolated and fashioned into marketable form, it must be moved across space – from seller to buyer – for its value to be realized, that is, for profit to be made. In short, mineral commodities "become" (see Adey 2006) through the instigation and management of a diverse assortment of human and non-human movements.[1]

But mining is not all fluxes and flows: it also requires what David Harvey (1996, 81) refers to as "permanences" in the social and material world – including permanences that *facilitate* mining's movements. In developing his relational theory of space, place, and environment, Harvey urges us to think about permanences as the crystallization of social processes that achieve relative stability – *relative* because "no matter how solid they may seem – [permanences] are not eternal" (Harvey 1996, 261). The permanences on which mining's movements depend vary according to the type of mineral extracted, the form and scale of the extraction process (e.g., large-scale, capital-intensive versus small-scale, labor-intensive), and the broader socio-natural context in which the activity takes place. They may include formal and informal institutional frameworks, such as property regimes, routinized divisions of labor, and trade agreements. They may also include physical infrastructure: things like chemical plants, deep-water ports, fiber-optic cabling,

DOI: 10.4324/9781003313236-1

housing complexes, road networks, and slurry pipelines. To the extent that such forms of infrastructure are "fixed" in the landscape, the movements and mobilizations that constitute mining are made possible by – and exist in dialectical relationship to – what Peter Adey (2006, 83) refers to as "relative immobilities" (see also Dessertine et al., Chapter 7 this volume). Attention to these infrastructures also serves as a reminder that mining's movements are "always located and materialized" (Sheller and Urry 2006, 210).

Mining also puts into motion people and things that escape control (Bebbington and Humphreys Bebbington 2018). As toxic substances used in or generated through mining and metallurgical processes are released into the environment, they may travel airborne or through the movement of water into bodies and ecologies far removed from the mining site (Bridge 2004; Kemp et al. 2010; Perreault 2013). The accumulation of mining rents presents opportunities for corruption, or the redirection of flows of public money into private accounts (Dougherty 2015). Driven by the prospect of stable employment, people may flock to newly initiated mining projects in numbers that mining companies cannot – or would rather not – absorb (Banks and Schwörer, Chapter 8 this volume). Or people may decide to travel to mining areas in pursuit of other economic opportunities, like the chance to share in compensation payments (e.g., by claiming connections of kinship or other forms of relatedness; see Bainton and Banks 2018), to engage in "informal," or small-scale, mining (Bainton et al. 2020; Bryceson, Jønsson, and Shand 2020; Dessertine et al., Chapter 7 this volume), or to establish local businesses (Banks and Schwörer, Chapter 8 this volume; Mususa and Peša, Chapter 6 this volume). Unfulfilled promises of mining-led local development can spur residents to organize and protest, including by blocking the flow of materials into and out of the mining site (Himley 2013). These social and material "overflows" (see Le Meur, Chapter 9 this volume) often themselves become the subject of formal or informal governance mechanisms aimed at, for instance, capturing a portion of profits (as rent) from mobile small-scale miners (see Dessertine et al., Chapter 7 this volume) or mollifying those protesting large-scale operations through compensation schemes or firm-led social development programs (see Damonte, Godfrid, and López, Chapter 3 this volume; also, Frederiksen and Himley 2020).

These examples of overflows are also a reminder of the need to account for and understand the agency of people – and of things (see Dunia Kabunga, Marijsse, and Geenen, Chapter 5 this volume) – in shaping the nature and form of the movements triggered by, required for, or otherwise related to mining. The people that mining puts into motion – purposefully or not – have desires, needs, and senses-of-self that may be shaped by the logics and requisites of extractive economies, but are never fully determined by these (Verweijen, Geenen, and Bashizi 2022; Zhu and Klein 2022). In broad terms, mining-related population movements can thus be thought of as "neither perfectly ordered nor anarchic" (Sheller and Urry 2006, 216): they are the products of a complex mixture of multi-scalar political-economic forces *as well as* the agency of those individuals and groups in motion. Though, the degree to which people have exercised autonomous agency in their mining-related movements – and, conversely, the extent to which these movements have been directed by powerful actors and institutions – has varied markedly across time and space, as well as across sectors of the mining economy (see below).

This edited volume centers the concept of *mobility* to explore the power relations that shape how and why people move across space in relation to mining, as well as the multiple dimensions and effects of these movements.[2] As Mimi Sheller (2018, 18) explores, the past two decades have seen the emergence within the social sciences – as well as related applied fields like design and urban planning – of a new interdisciplinary approach to the study of mobilities (see also Adey 2006; Cresswell 2010a, 2012, 2014; Sheller and Urry 2006). This proliferation of mobility studies has occurred through, *inter alia*, the organization of associations, conferences, and working groups; the publication of monographs and edited volumes (e.g., Adey et al. 2014; Elliott and Urry 2010; Scheller and Urry 2006; Urry 2007); and the establishment of two journals, *Mobilities* and *Transfers: Interdisciplinary Journal of Mobility Studies*. It is impossible to neatly characterize this wide-ranging and dynamic body of scholarship that addresses multiple forms of movement. In broad terms, however, mobilities research underscores the "constitutive role of movement" in social life and aims to understand "the organization of power around systems of governing mobility and immobility at various scales" (Sheller 2018, 19). In this sense, work adopting a mobilities approach places itself in opposition to "sedentarist" epistemologies and approaches that assume fixity and boundedness to be the norm, and movement a deviation from that norm (Adey 2006; Sheller and Urry 2006). In doing so, mobilities research highlights the need to understand the relationships of human (im)mobilities to the (im)mobilities of non-human things (capital, information, objects, etc.), as well as the discourses, meanings, and representations that imbue and frame mobile lives and phenomena (Cresswell 2010b; Sheller 2018). This work also stresses the importance of interrogating the politics of mobility, broadly construed to encompass how forms of social differentiation and hierarchy – for instance, along lines of caste, citizenship status, class, gender, physical ability, race, and sexual identity – shape accessibility to and experiences of movement, in the process generating "uneven" or "differentiated" mobilities (see Adey 2006; Sheller 2018).

How are mining-related mobilities shaped by the historical-geographical contexts and socio-political arrangements in which they take place? What meanings do people attach to their own mining-related movements or to those of others? How do the movements of people vis-à-vis mining relate to the movements of other material and non-material things? How do mobility practices inform processes of social differentiation in mining regions? How can an analytical focus on spatial mobility aid scholars and policy makers in understanding the complex relations between mining and social change? These questions lie at the core of this volume. To address them, the volume brings together research undertaken in three world regions with significant histories of mineral extraction and where mining currently plays an important role in socio-economic life: the Andes, Central and West Africa, and Melanesia.

In the next section, we offer a historical perspective on the mining-mobility nexus and identify key themes addressed in this volume. In the final section, we situate the volume within the social-science literature on mining and outline the contents of individual chapters.

Mining-mobility linkages in historical context

Researchers underscore the importance of historical perspectives in mobilities scholarship, including to reduce the risk of presenting mobilities as somehow "new" – i.e., the product of "late-capitalist" technologies such as high-speed Internet, mobile phones, and jetliners (Cresswell 2012, 2014). In the case of mining, such technologies, in addition to being constructed of mineral commodities, have certainly shaped forms, patterns, and velocities of human and non-human movement vis-à-vis mining across small-, medium-, and large-scale sectors, one example being the development of "fly-in/fly-out" commute work arrangements (Storey 2010). A historical perspective, nonetheless, remains vital. Mining has long been linked to the movement of people across space (Ferguson 1999; Moodie and Ndatshe 1994; Yusoff 2019) and these historical processes of mobility and migration continue to imprint social and political life in mining regions today, as several chapters in the volume explore.

Two key mining-sector dynamics underlie the historical linkages between mineral extraction and human spatial mobility. On the one hand, mining has exhibited across time a "profound geographical restlessness" (Moore 2007, 129–130), with spatial expansion into "new ground" a principal strategy for maintaining or increasing production. This restlessness can be witnessed at multiple spatial and temporal scales. At the world-historical scale, for instance, it is seen in the relocation of silver mining, across the sixteenth and seventeenth centuries, from central Europe to the central Andes (present-day Bolivia and Peru) and then to New Spain (present-day Mexico), following a logic of what Jason Moore (2007, 125), drawing on Madhav Gadgil and Ramachandra Guha (1992), refers to as "sequential overexploitation." It is also seen, albeit at different spatial and temporal scales, in the rapid emergence – and, not uncommonly, abrupt decline – of artisanal or small-scale gold mining sites, as has occurred in recent decades in countries as diverse as Guinea, Indonesia, Mongolia, Peru, and Tanzania (Bryceson 2018; Bryceson, Jønsson, and Shand 2020; Damonte 2018; Dessertine et al., Chapter 7 this volume; Dunia Kabunga, Marijsse, and Geenen, Chapter 5 this volume; Zhu and Peluso 2021).

On the other hand, relations between mining and human mobility have been strongly shaped by the sector's labor requirements, both for extraction itself and to produce necessary inputs. The spaces into which mining activities expand – i.e., "mining frontiers" – are often relatively remote or peripheral areas without a large pool of potential workers (Ciccantell and Gellert 2022). As a result, the expansion of the mining frontier has traditionally been accompanied by a process of in-migration. Sometimes, this movement of people has been managed by powerful actors, such as colonial states (see, for instance, Dunia Kabunga, Marijsse, and Geenen, Chapter 5 this volume; Le Meur, Chapter 9 this volume; Marston, Chapter 2 this volume). At other times, these labor mobilities are more spontaneous and less "managed," as has often been the case with historical and contemporary gold rushes involving "individualized, highly mobile artisanal gold miners" (Bryceson 2018, 32; Zhu and Peluso 2021). Meanwhile, as Paul S. Ciccantell and Paul K. Gellert (2022, 42) discuss, the transition from an extractive frontier to what they refer to as an "extractive periphery" typically involves "the establishment of a more permanent pool of exploitable (relatively cheap) labor," while the decline of mineral extraction activities (as a result, for instance, of resource exhaustion) often leads to processes of out-migration.

Many chapters in this volume contextualize their studies of recent mining-mobility dynamics within longer histories of region-specific population movements. Here, we briefly review two historical examples: silver mining in colonial Potosí and the California gold rush. Our aim is not to provide thorough accounts of these iconic cases in the global history of mining; expansive literatures exist on both. Rather, we leverage these examples to illustrate the general dynamics that have underlain mining-mobility relations, to identify important differences in the human mobilities that mineral extraction rested upon and stimulated historically, and to introduce key themes addressed by contributors to this volume.

Silver mining in colonial Potosí. From its earliest days, Spanish colonization of the Americas was driven by a restless search for precious metals, namely gold (Bury and Bebbington 2013; Galeano 1971). In the Central Andes, once opportunities to plunder already mined gold from Indigenous societies declined, the colonial administration turned its attention to silver, especially after a major deposit was identified in 1545 at a site that would become known as Potosí, in what is now Bolivia. In the context of Indigenous demographic collapse (Cook 2010) – due to disease, warfare, and other factors – Spanish colonialists faced an acute need to secure an ample and consistent labor force for mining, metallurgy, and other economic pursuits, like agriculture and textile manufacturing. Under the direction of Viceroy Toledo, the Crown's response was the *mita*, an adapted version of the Incan system of temporary compulsory labor (Dell 2010; Marston, Chapter 2 this volume; Moore 2007; Robins 2011). Under the *mita*, starting in 1573, Indigenous communities within a defined geographical area were required to provide, on a rotational basis, one-seventh of their adult male population to work in the mines and mills of Potosí.[3] This was coupled with a related Toledo-initiated effort of forced Indigenous resettlement into nucleated villages, known as *reducciones*, which, as Heidi V. Scott (2009, 69) contends, "reflected a powerful desire to arrest uncontrolled indigenous mobility and to make native populations visible and easily accessible." Together, the *reducciones* and the *mita* – the latter of which operated for more than two centuries until its abolition in 1812 – had major, long-term implications for population geographies and mobility dynamics in the Central Andes (Castillo Guzmán 2020; Dell 2010; Moore 2007; Scott 2009). For one, many *mita* conscripts, known as *mitayos*, died due to the dangerous conditions they faced on their journeys to the mines, along with the hazardous working conditions in the mines themselves, thus contributing to both ongoing Indigenous demographic decline (Dell 2010; Stavig 2000) and the crafting of proposals for how to address it (Uparela 2022). Additionally, while the *mita* was designed to facilitate and direct the circulation of laborers between their communities of origin and the mines, it also had the effect of sparking diverse "uncontrolled" mobilities. This is perhaps most clearly seen in the movements of *forasteros*, or individuals who fled their home communities to escape service, a phenomenon that vexed colonial administrators, depleted the populations of communities subject to the *mita*, and tended to place additional labor burdens on those who remained (Glave 1989; Robins 2011; Scott 2009).

California gold rush. The spatio-temporal restlessness of mining is also on display in the case of the California gold rush, a multi-year event sparked by the discovery of gold at a site known as Sutter's Mill on the western slope of the Sierra Nevada mountain range, in January of 1848. As with other major nineteenth-century gold rushes – such as those in Victoria (Australia), the Witwatersrand (South Africa), and the Klondike (Canada) – a defining feature of the California gold rush was "an avalanche of local in-migration" (Bryceson 2018, 34). By 1854, an estimated 300,000 people, mostly men, had migrated to California – which Mexico had ceded to the United States in February 1848 as part of the Treaty of Guadalupe Hidalgo – with many traveling thousands of miles across land and sea, impelled by dreams of wealth and adventure (Bryceson 2018; Rohrbough 2000; Taniguchi 2000).[4] Those descending upon California were of diverse geographical origin: while many migrants came from other parts of the United States, large numbers came from abroad, including, according to Sucheng Chan (2000, 44), "from various countries in Europe, as well as from Mexico, Chile, Peru, Canada, Hawaii (at that time an independent kingdom), China, Australia, and New Zealand." This mass migration to the California gold fields was facilitated by the form of mining that predominated, especially during the early years of the rush – that is, the labor-intensive exploitation of readily accessible streambed deposits, known as placers, with the aid of relatively simple technologies (Isenberg 2005; Rohrbough 2000). Coupled with the low barriers to entry that characterized this form of mining was an emergent ideology that framed gold as a "free good easily available in nature" (Isenberg 2005, 23), and the search for it in California "the epitome of American economic democracy" (Rohrbough 2000, 28). Importantly, as we discuss below, the reality of the California gold rush did not live up to these lofty ideals: with the gold fields marked by racism, nativism, and violence against marginalized groups, including Indigenous peoples (Chan 2000; Shaler 2020), the "economic democracy" of the rush was in many ways "delimited to men of European descent" (Bryceson 2018, 41). Moreover, the relatively accessible placer mining that initially characterized the California gold rush gave way within several years to more industrial and capital-intensive forms of mining, especially hydraulic mining. As Andrew Isenberg (2005, 24) notes, this shift in the techno-organizational form of gold mining "initiated the transformation of the gold country from a locale dominated by independent prospectors to an industrial place characterized by wage laborers."

These brief overviews of silver mining in colonial Potosí and the California gold rush suggest some important ways in which the human mobilities that have constituted and accompanied mineral extraction have varied historically. For one, there were significant differences in the *scale* of laborer movements in the two cases. The Potosí *mita* incorporated, as laborers, individuals who were from communities within a specific geographical catchment area in the Central Andes and who belonged to a defined ethno-legal category – "Indian," in the language of the day (Dell 2010; Stavig 2000). In contrast, the mass migrations of the California gold rush were decidedly more international in scope, with migrants traveling from various countries around the globe, in the process turning mid-nineteenth-century California into – according to Malcolm Rohrbough (2000, 26) – "the most cosmopolitan place in the world."[5]

In addition to differences in scale, the labor mobilities of the Potosí *mita* and California gold rush varied in *rhythm* and *temporality*. The *mita* was designed by the Spanish colonial administration to foster the temporary, circular migration of workers, and the system was in operation for more than two centuries (Robins 2011). The comparatively short-lived California gold rush, in contrast, was defined by an intense burst of in-migration. This was followed, after several years, by a transition in the method and organization of gold mining – that is, toward capital- and technology-intensive hydraulic mining – and a concomitant decline in independent prospecting (Isenberg 2005). While some gold-rush miners stayed in the mining sector as wage labors, many left the gold fields – some returning home, others moving on to new mining sites (e.g., to Australia, where a gold rush commenced in 1851), and many staying in California to work in other sectors of the state's economy, such as agriculture (Bryceson 2018; Chan 2000).

Additionally, the *systems of power* through which mining-related mobilities were governed were markedly distinct in the two cases, reflecting in many ways the distinct political-economic regimes in which the two took place. With the *mita*, conscripts from Indigenous communities were sent to Potosí as part of a highly structured, forced labor draft. In comparison, the migrants of the California gold rush tended to be more self-directed, their movements less ordered by centralized governing institutions. There are important caveats to these depictions. Social histories of the *mita*, for instance, suggest ways that Andean peoples subject to the draft exercised agency, beyond the many cases of individuals evading service through flight. For example, after completing their labor obligations, some *mitayos* remained in Potosí or traveled to other mining sites to work as wage laborers (Robins 2011). In the California gold rush, some individuals arriving to the gold fields were not "self-propelled" (Bryceson 2018, 32) adventure-seekers but enslaved African Americans accompanying their owners, while others arrived as bonded laborers (Chan 2000). Further, laws and government policies clearly shaped the movements and experiences of gold-rush participants, one example being a Foreign Miners' Tax, primarily leveled on non-English-speaking miners, that served to "harden" racial and ethnic divisions and stoke tensions between miners of different national origins (Chan 2000; Shaler 2020).[6] In sum, while there were broad and important differences in the power-geometries (see Massey 1991) that shaped and constrained mining-related mobilities in the two cases, these systems of power were complex and uneven in their own distinct ways, and with concrete human consequences.

Matters of scale, rhythm/temporality, and power/governance remain central to studies of more contemporary mining-mobility dynamics, as chapters in this volume illustrate. Several other themes present in both the Potosí *mita* and California gold rush cases also receive focused attention in the case studies that authors analyze in this book.

One is the relationship between social differentiation – for instance, along lines of ethnicity, gender, and race – and mining-related (im)mobilities. In both historical case studies, forms of social difference and hierarchy were central to the organization of the mining labor force – and, thus, to patterns of mining-related movement across space. Histories of these cases further suggest ways that mining-related (im)mobilites were, in turn, *productive* of social differentiation and identity formation – one example being nativists laws, like the Foreign Miner's Tax, which structured access to California gold fields along lines of national origin (Chan 2000). Several chapters in this volume address the intersections of mining, mobility, and social differentiation in various contemporary contexts, including through analyses of the relations between small-scale mining mobilities and Indigenous autonomy in Bolivia (Chapter 2 by Andrea Marston), the gendered dimensions of new forms of spatial mobility linked to large-scale mineral development in Peru (Chapter 4 by Gerardo Castillo Guzmán), and the lingering influences of colonial policies of spatial and racial segregation on mining mobilities and place-based attachments in New Caledonia (Chapter 9 by Pierre-Yves Le Meur).

Second, the historical cases point to the need to account for how changes in the technological and organizational form of mining, such as the development of mercury amalgamation in the case of Potosí and the shift to hydraulic mining in California, along with associated shifts in modes of governance, have shaped mining-related mobility dynamics. This is an issue that numerous chapters in the volume also take up, including through studies of how the evolution of accommodation arrangements, understood as a kind of infrastructure, have shaped mobilities in the context of large-scale mining (LSM) in Chile and Peru (Chapter 3 by Gerardo Damonte, Julieta Godfrid, and Ana Paula López) and of the interrelated movements of people, technologies, and materials across different time periods in the gold fields of Eastern Democratic Republic of the Congo (DRC) (Chapter 5 by Philippe Dunia Kabunga, Simon Marijsse, and Sara Geenen). Relatedly, case studies in the volume focus attention on how the nature and rhythms of mining-related mobilities change over time in relation to the mineral-project life cycle and the distinct labor requirements of each stage – as in Glenn Banks and Tobias Schwörer's study of migration in the context of LSM projects at different stages of development in Papua New Guinea (Chapter 8), as well as in Séverine Bouard and Valentine Boudjema's study of workforce mobility and livelihoods in the North Province of New Caledonia (Chapter 10).

Third, the Potosí *mita* and California gold rush cases put on display the role of mining mobilities in the making of place, a term used in human geography and cognate fields to broadly refer to not just the physical characteristics of a locale but the meanings and significance that diversely situated groups and individuals attach to these (Bebbington and Humphreys Bebbington 2018; Cresswell 2004; Macintyre 2018). In both cases, mobilities put in motion by mining had significant impacts on the character of places impacted, including "sending" and "receiving" locales. In this volume, several chapters address the relationship between mining and place-making through a focus on mobility. These include Patience Mususa and Iva Peša's chapter on the diverse forms of human mobility that contribute to the "making" of mining localities in the Zambian Copperbelt (Chapter 6), as well as Anna Dessertine, Robin Petit-Roulet, Muriel Champy, and Ibrahima Kalil Doumbouya's chapter on the relations between artisanal and small-scale gold mining (ASGM) mobilities and localized, place-based forms of territorial governance in Guinea and Côte d'Ivoire (Chapter 7).

Situating this volume and introducing its chapters

As the previous section underscored, mining has long been interlinked with human mobility, in diverse forms. Much late-twentieth-century mining-focused social-science scholarship indeed focused on the relationship between mining and population movements, typically understood through the concept of "migration" (see, *inter alia*, Cleary 1990; Ferguson 1999; Moodie and Ndatshe 1994). In recent decades, however, coinciding with a general expansion of academic interest in the socio-environmental dimensions of extractive industries, there has been a shift in emphasis toward more "territorial" approaches. Here, rather than "following the people"[7] who are put in motion by mining activities, the implications of and struggles over mining tend to be studied within the context of geographically defined areas, such as the bounded territories of one or more specific communities, generally located in the mine's immediate surroundings (for instance, Arellano-Yanguas 2017; Bury 2008; Himley 2016; Hinojosa 2013; Larmer 2021; Loayza and Rigolini 2016; Ticci and Escobal 2014). This analytical approach, it bears mentioning, in many ways parallels the territorially based method of delineating an operation's "area of influence" that prevails within the global mining industry.

An important departure from this trend is research on artisanal and small-scale mining (ASM), in which the movement of miners is – or has remained – a key emphasis (for example, Bryceson, Jønsson, and Shand 2020; Dessertine 2016; Jønsson and Bryceson 2009). It should also be noted that within scholarship on LSM, including research adopting a territorial approach, population movements are often critical to the dynamics being studied, even if questions of migration and mobility are not foregrounded analytically. This is the case, for instance, with research on the socio-environmental dimensions of displacement and resettlement required for mine development or expansion (Bury 2007; Owen and Kemp 2015; Szablowski 2002), on the secondary or indirect forms of displacement that result from the influx of people into new mining areas, and the pressure on land and resources that this generates (Bainton and Banks 2018; Bainton, Burton, and Owen 2021), and on processes of urbanization related to mining and other extractive industries (Fafchamps, Koelle, and Shilpi 2017; Jønsson and Bryceson 2017; Udelsmann Rodrigues et al. 2021). In this context, an emergent body of work has (re)engaged questions of population movement vis-à-vis mining (small- and large-scale) through an explicit conceptual engagement with the "new mobilities" framework discussed in the introduction to this chapter (e.g., Bolay 2022). It is within this emergent body of theoretically informed work on the mining-mobility nexus that we situate the present volume.

Interest in advancing this book project was also cultivated through the participation of two of the editors (David Brereton and Gerardo Castillo Guzmán) in a 2017 symposium, hosted by the Centre for Social Responsibility in Mining at the University of Queensland, entitled "Human migration, development and the global mining industry." Papers originally presented at this symposium – including one by Castillo Guzmán and Brereton (see Castillo and Brereton 2018) – were subsequently published in a special issue of *Sustainable Development*, entitled "Mining, mobility and sustainable development" (Bainton, Owen, and Kemp 2018). In ensuing years, our collective interest in further pursuing questions of mining and mobility grew, and our motivation to pursue an edited volume centered on three world regions derived from a desire to explore both how mining-mobility connections are shaped by the specific historical-geographical characteristics of the regions in questions and what comparison across regions can reveal about the general processes and dynamics at play. In selecting the Andes, Central and West Africa, and Melanesia as our three case-study regions, we recognized some broad similarities, including that mining has long been a driver of population movements and territorial transformations in the three regions, and all three have garnered significant interest from scholars of the socio-environmental dimensions of mining. At the same time, their different environmental, political-economic, and social histories, along with distinct colonial experiences and historical relations to extraction, make the three regions especially suitable for comparative analysis.

In recruiting authors for the project, we aimed to assemble an interdisciplinary group of scholars (including as a means of encouraging dialogue across disciplinary divides), and we prioritized potential contributors with substantial research experience focused on their respective region. On both counts we were successful: each chapter in the volume is written by an author or authors who have conducted extensive research on – and thus have deep knowledge of – their case-study area(s), and collectively the authors have expertise in a range of social-science fields, including anthropology, development studies, history, geography, international development, and sociology. To foster productive exchanges among volume contributors, we held two virtual workshops for contributors, the first in April of 2022 and the second in October of 2022. The workshops provided opportunities for authors to exchange ideas and feedback on draft chapters at early and advanced stages. They also allowed us to identify both cross-cutting themes and divergences across the different case-study chapters, which we further explore in the volume's concluding chapter.

Before providing more details on each of the individual chapters in the volume, it is worth highlighting that while most chapters focus on what can broadly be termed LSM – a capital-intensive, highly mechanized form of mining often associated with a large environmental footprint – the book also includes several case studies focused on ASM. This is an important sector of the mining industry, which provides livelihoods for millions of people globally and extends over relatively large areas in some countries – indeed, even if the social, economic, and environmental impacts of individual miners and groups of miners may be relatively small, their collective footprint often is not. The relations between LSM and ASM can be complex, and sometimes conflictual, such as in cases in which the two sectors compete over the same deposits (see Libassi 2022). Yet, as chapters in the volume show, the ASM sector is often characterized by distinctive mobility dynamics that warrant ongoing analysis.

The Andes

Chapter 2, by Andrea Marston, is the first of the volume's three-chapter section on the Andes. In this chapter, Marston adopts a mobilities framework to analyze the relationship between small-scale tin mining and Indigenous autonomy in Norte Potosí, Bolivia. As our discussion of the Potosí *mita* above suggested, Indigenous peoples in this part of the Central Andes have a long history of engagement with extractive economies, though Marston documents that during most of the twentieth century few workers in the tin mines of Norte Potosí were from surrounding Indigenous communities. This situation started to change with the establishment of mining cooperatives in the mid-1980s. At this historical juncture – which coincided with Bolivia's late-twentieth-century transition to neoliberalism – an increasing number of people from rural areas of Norte Potosí migrated to work in the region's mines, while also maintaining connections of multiple types with their home communities. According to Marston, the result has been two new kinds of mobility: that of Indigenous people between the mines and their home communities and that of money (mining-derived profits) within rural areas, where such income is increasingly viewed as a means of supporting Indigenous autonomy. To conceptualize these complex dynamics, Marston, drawing on Silvia Rivera Cusicanqui (2018), mobilizes the Aymara notion of *ch'ixi* (motley), broadly understood to refer to the "non-synthetic" mixture – and coexistence – of Indigenous and non-Indigenous worlds.

In Chapter 3, Gerardo Damonte, Julieta Godfrid, and Ana Paula López train their sights on infrastructure and its role in shaping patterns of everyday spatial mobility in the context of LSM. Adopting a comparative approach, the chapter draws on research undertaken in Espinar, Peru, and Los Andes, Chile, where the Tintaya-Antapaccay and Andina copper mines are located, respectively. Damonte, Godfrid, and López are especially interested in accommodation arrangements for workers and displaced people, and they trace how broad changes in state and corporate approaches to the provisioning of this kind of infrastructure have shaped local mobility practices. They draw attention to a shift away from the construction of spatially concentrated accommodation arrangements (e.g., mining camps and towns built by companies, typically adjacent to mines) and toward the establishment of housing and other accommodation infrastructures in cities and mining suburbs. The chapter ties these transformations in the form and geography of accommodation arrangements to shifting production systems and governance regimes, and it finds that the results, for workers and displaced people, are more dispersed and spatially expansive patterns of everyday mobility.

The Andes section of the volume concludes with a chapter by Gerardo Castillo Guzmán (Chapter 4) examining the implications of LSM development for women's spatial mobility and autonomy. Castillo Guzmán's case study is the La Granja copper project, in northern Peru, where decades of exploration and development activities have had significant impacts on local society, even though mining has yet to commence (and the project is currently on hold). Based on in-depth qualitative research, Castillo Guzmán finds that while the impacts of mineral-resource development on the social, political, and economic activities of women from La Granja have been complex, women have experienced an increase in their capacity to move freely across space, though mobility limitations rooted in the area's patriarchal social structure persist. Through analysis of this case, Castillo Guzmán aims to build a bridge between the literature on gender and extractive industry and research on the mining-mobility relationship. In the process, the chapter underscores the importance of considering how social divisions and hierarchies – related to not only gender but also age, ethnicity, race, etc. – influence the capacity of individuals and groups to navigate landscapes of mineral extraction.

Central and West Africa

The next three chapters comprise the Central and West Africa section of the volume. In Chapter 5, Philippe Dunia Kabunga, Simon Marijsse, and Sara Geenen draw on extensive ethnographic research on ASGM in Eastern DRC to analyze different kinds of mobility – including that of mine workers, mining technologies, and materials – in and around the gold mines. In the DRC, both industrial and artisanal mining have sparked major population movements across time, whether as the result of companies' labor recruitment strategies during the colonial period or through rural people leaving their villages to work in ASGM today. However, as Dunia Kabunga, Marijsse, and Geenen assert, not only people are on the move in and around the gold mines of Eastern DRC: so are capital, knowledge, skills, and technologies; and so is gold itself. The chapter mobilizes the concept of "resource assemblage" to analyze the multiple forms of movement that come together in the extraction of gold, and through close attention to miners' framings of gold's movements, underscores the need to address questions of an ontological nature about the agentic qualities of subterranean materials.

In Chapter 6, Patience Mususa and Iva Peša investigate the linkages among mining, mobility, and place-making in the context of large-scale copper mining in Zambia. The authors focus on three case studies, comprising two older mining districts and one newer one, and draw on a mix of quantitative and qualitative data sources – national censuses, a population survey, life history interviews – to add fresh insights to a rich history of scholarship on migration and Zambian mining localities. Mususa and Peša argue that while mining has certainly shaped mobility patterns in the case study areas, these patterns have been influenced by a range of additional factors, including the chance to participate in livelihood opportunities (e.g., agriculture, forestry, trade) and through individuals' social ties and religious connections. These findings call into question the notion that mining has been *the* driver of population movements in the region, while also pointing to the importance of understanding the varied ways that extraction interacts with context-specific social and economic processes to shape understandings of place in mining localities. Tracing people's complex mobilities in Zambia's copper-mining localities further leads Mususa and Peša to question the characterization of these localities as "mining enclaves." As an alternative, they prefer the concept of "mining hubs" as a way of recognizing the diverse linkages that urban mining localities have long maintained with other areas, including rural hinterlands.

The focus returns to ASGM in Chapter 7. Drawing on ethnographic research carried out in gold mining areas in Guinea and Côte d'Ivoire, Anna Dessertine, Robin Petit-Roulet, Muriel Champy, and Ibrahima Kalil Doumbouya analyze the interactions between ASGM mobilities and territorial governance, while also assessing the impacts of ASGM mechanization on these interactions. The chapter documents how people engage in diverse kinds of ASGM mobilities in Guinea and Côte d'Ivoire (e.g., short- vs. long-term; autonomous vs. supervised), with flows of goods, information, people, and technologies linking mining sites into networks that assume the form of an "archipelago." The authors show that while local-level governance mechanisms play a role in regulating ASGM in both countries, the nature of these mechanisms varies, with those in Guinea being more collective and those in Côte d'Ivoire more individualized. These differences hold significant implications for mining-mobility trajectories. The authors also demonstrate that ASGM mechanization, through its impacts on the geographies and social relations of gold production, is influencing mobility patterns and triggering reconfigurations of governance frameworks. Through their nuanced analysis of the complex interactions between mobility, governance, and technological change, Dessertine et al. underscore the importance of attending to the agency of mobile miners as well as the relatively immobile – i.e., territorially based – systems of authority and power that enable and control miners' activities.

Melanesia

In Chapter 8, the first of three chapters on Melanesia, Glenn Banks and Tobias Schwörer focus on migration patterns in the context of four sites of LSM activity in Papua New Guinea: Frieda River, Wafi-Golpu, Porgera, and Lihir. These cases provide rich opportunities for comparison; in part because of their distinct geographical and socio-economic settings, and because two are projects still at the advanced exploration stage, whereas two are mines that have been operating for decades. With the aim of refining a model that Nicholas A. Bainton and Banks (2018) developed to represent the interrelations between migration, social processes, and landscapes in the context of mining development, Banks and Schwörer explore how the regional context and the specific forms of social organization found at each site have affected the nature and scale of mining-related migration. Especially significant are the authors' findings regarding the different notions of kinship and relatedness among area social groups, and how these have enabled and regulated in-migration to the different sites to varied degrees and in diverse ways. Although there are major differences in geographical and sectoral focus, the analysis by Banks and Schwörer overlaps with that of Dessertine et al. in Chapter 7, as both chapters highlight the significance of local forms of socio-political organization in shaping mining mobilities.

Chapter 9, by Pierre-Yves Le Meur, explores the role of mobility in the historical construction of New Caledonia as a single yet socio-spatially heterogeneous mining enclave. Le Meur documents four waves of mining-related immigration to New Caledonia, the first dating to the late nineteenth century, in the process tracing complex historical linkages between the settler colonial project, racial segregation, and mining. The focus of the chapter then turns to the locality of Thio, a long-standing nickel-mining center in the South Province, which is representative of the varied forms of mining-related mobilities found in New Caledonia during different historical periods. Through discussions of three brief case studies – all with an emphasis on relations between mining and Kanak communities and political struggles – Le Meur examines how the destructive/productive processes constituting mining have shaped residential patterns, senses of belonging, and place-based attachments in Thio. Le Meur thus illustrates, through the case of Thio, the dialectical relation between mobility and locality, while also highlighting the need to account for not just human flows but also non-human ones (e.g., those of contaminants and money), when analyzing the nexus of mining and mobility.

Staying with New Caledonia, Chapter 10, by Séverine Bouard and Valentine Boudjema, documents and analyzes the circulation of employees of the Koniambo Nickel SAS (KNS) mining and metallurgical company in New Caledonia's North Province. Long-term research in the area, involving a mix of qualitative and quantitative methods, allows Bouard and Boudjema to grasp changes in employee mobility patterns across the different stages of the mining life cycle, to capture material and non-material dimensions of employee movements, and to assess the territorial and livelihood impacts of these flows. The chapter focuses attention on the circulation of Kanak individuals between their traditional territories, where mining activities are absent, and the mining territory where KNS operates. This reveals important findings about not just how mining materially and symbolically impacts Kanak lives but also how Kanak individuals have utilized mining employment to acquire different forms of human, physical, and social capital, in the process building livelihood opportunities in – and maintaining connections to – non-mining territories. Through their multi-scalar approach, Bouard and Boudjema – like Mususa and Peša in Chapter 6 – question the utility and value of traditional conceptualizations of "mining enclave."

Conclusion

Finally, in Chapter 11, we, the editors, offer a synthesis of key findings emerging from the analyses presented in the volume, while also commenting on governance implications of these findings and identifying priorities for future research on the mining-mobility nexus. We focus particularly on human mobility, organizing our discussion of the volume's key insights around a typology that includes inward flows, circular and itinerant flows, local flows, and outward flows, and connecting findings from the volume's case studies with additional academic scholarship on mining and mobility. We identify *complexity* and *human agency* as important overarching themes, and we discuss how attention to these can productively inform efforts to address governance challenges vis-à-vis mobility in both LSM and ASM sectors. We conclude with four suggestions for extending research in this vibrant and important area of scholarship, while also highlighting how expected changes in the mining landscape in upcoming decades will make research on the relations among mining, mobility, and social change even more essential.

Notes

1 We thank Tenley Banik and Margaret Himley for valuable input while we drafted this chapter.

2 By "space" we refer to not just physical space but also social space, as different spaces typically entail distinct cultural, economic, political, and social "logics" – for instance, those of subsistence farming vs. paid mining work or those of kinship-based relations vs. corporate policy.

3 *Mita* laborers were also conscripted to work the mercury mines of Huancavelica (in present-day Peru), which assumed strategic importance for the colonial administration after the adoption of mercury amalgamation – a processes involving the use of mercury to separate silver from finely ground ore – at the Potosí mines in the 1570s (Moore 2007; Robins 2011).

4 Nancy J. Taniguchi (2000) provides an account of the diverse experiences of women in California during the gold rush and the ways in which the rush transformed their lives.

5 In addition to people of Indigenous and European descent, early colonial Potosí was home to a significant number of enslaved Africans, though relatively few worked underground in the mines, but rather labored in activities including blacksmithing, domestic service, llama and mule driving, and silver refining (Lane 2019).

6 The inequities and injustices of the California gold rush were experienced especially acutely by the Indigenous people of California. While Native Californians made up a significant percentage of the mining labor force in the early months of the gold rush, and early migrants relied on Indigenous knowledge of gold regions (Shaler 2020), they were soon displaced by immigrant miners, who "engulfed Native American land" (Bryceson 2018, 41), and relations between settlers and Indigenous Californians came to be characterized by "catastrophic violence" (Shaler 2020, 80).

7 We thank Andrea Marston for this phrase.

References

Adey, Peter. 2006. "If Mobility Is Everything Then It Is Nothing: Towards a Relational Politics of (Im)mobilities." *Mobilities* 1 (1): 75–94. doi: 10.1080/17450100500489080.

Adey, Peter, David Bissell, Kevin Hannam, Peter Merriman, and Mimi Sheller. 2014. *The Routledge Handbook of Mobilities*. New York: Routledge.

Arellano-Yanguas, Javier. 2017. "Inequalities in Mining and Oil Regions of Andean Countries." *Iberoamerican Journal of Development Studies* 6 (2): 98–122. doi: 10.26754/ojs_ried/ijds.255.

Bainton, Nicholas A., and Glenn Banks. 2018. "Land and Access: A Framework for Analysing Mining, Migration and Development in Melanesia." *Sustainable Development* 26 (5): 450–460. doi: 10.1002/sd.1890.

Bainton, Nicholas A., John R. Owen, and Deanna Kemp. 2018. "Mining, Mobility and Sustainable Development: An Introduction." *Sustainable Development* 26 (5): 437–440. doi: 10.1002/sd.1889.

Bainton, Nicholas, John R. Owen, Simon Kenema, and John Burton. 2020. "Land, Labour and Capital: Small and Large-Scale Miners in Papua New Guinea." *Resources Policy* 68: 101805. doi: 10.1016/j.resourpol.2020.101805.

Bainton, Nicholas, John Burton, and John R. Owen. 2021. "Land Relations, Resource Extraction and Displacement Effects in Island Papua New Guinea." *The Journal of Peasant Studies* 49 (6): 1295–1315. doi: 10.1080/03066150.2021.1928086.

Bebbington, Anthony, and Denise Humphreys Bebbington. 2018 "Mining, Movements and Sustainable Development: Concepts for a Framework." *Sustainable Development* 26 (5): 441–449. doi: 10.1002/sd.1888.

Bolay, Matthieu. 2022. "Miners on the Move: Expellable Labor, Flexible Intimacies, and Institutional Recast in West African Artisinal Gold Mines." *American Ethnologist* 49 (4): 521–535. doi: 10.1111/amet.13108.

Bridge, Gavin. 2004. "Contested Terrain: Mining and the Environment." *Annual Review of Environment and Resources* 29: 205–259. doi: 10.1146/annurev.energy.28.011503.163434.

Bryceson, Deborah Fahy. 2018. "Artisanal Gold-Rush Mining and Frontier Democracy: Juxtaposing Experiences in America, Australia, Africa, and Asia." In *Between the Plough and the Pick: Informal, Artisanal and Small-Scale Mining in the Contemporary World*, edited by Kuntala Lahiri-Dutt, 31–61. Acton: ANU Press.

Bryceson, Deborah Fahy, Jesper Jønsson, and Mike Shand. 2020. "Mining Mobility and Settlement during an East African Gold Boom: Seeking Fortune and Accommodating Fate." *Mobilities* 15 (3): 446–463. doi: 10.1080/17450101.2020.1723879.

Bury, Jeffrey. 2007. "Mining Migrants: Transnational Mining and Migration Patterns in the Peruvian Andes." *The Professional Geographer* 59 (3): 378–389. doi: 10.1111/j.1467-9272.2007.00620.x.

Bury, Jeffrey. 2008. "Transnational Corporations and Livelihood Transformations in the Peruvian Andes: An Actor-Oriented Political Ecology." *Human Organization* 67 (3): 307–321. doi: 10.17730/humo.67.3.u8lv3g07w5711885.

Bury, Jeffrey, and Anthony Bebbington. 2013. "New Geographies of Extractive Industries in Latin America." In *Subterranean Struggles: New Dynamics of Mining, Oil, and Gas in Latin America*, edited by Anthony Bebbington and Jeffrey Bury, 27–66. Austin: University of Texas Press.

Castillo Guzmán, Gerardo. 2020. *Local Experiences of Mining in Peru: Social and Spatial Transformations in the Andes*. Abingdon: Routledge.

Castillo, Gerardo, and David Brereton. 2018. "Large-Scale Mining, Spatial Mobility, Place-Making and Development in the Peruvian Andes." *Sustainable Development* 26: 461–470. doi: 10.1002/sd.1891.

Chan, Sucheng. 2000. "A People of Exceptional Character: Ethnic Diversity, Nativism, and Racism in the California Gold Rush." *California History* 78 (2): 44–85. doi: 10.2307/25463688.

Ciccantell, Paul S., and Paul K. Gellert. 2022. "Migration, Resource Frontiers, and Extractive Peripheries: Toward a Typology." In *Migration, Racism and Labor Exploitation in the World-System*, edited by Denis O'Hearn and Paul S. Ciccantell, 29–44. New York: Routledge.

Cleary, David. 1990. *Anatomy of the Amazon Gold Rush*. London: Palgrave Macmillan.

Cook, Noble D. 2010. *La catástrofe demográfica andina: Perú 1520-1620*. Lima: PUCP.

Cresswell, Tim. 2004. *Place: A Short Introduction*. Oxford: Blackwell.

Cresswell, Tim. 2010a "Mobilities I: Catching Up." *Progress in Human Geography* 35 (1): 550–558. doi: 10.1177/0309132510383348.

Cresswell, Tim. 2010b. "Towards a Politics of Mobility." *Environment and Planning D: Society and Space* 28 (1): 17–31. doi: 10.1068/d11407.

Cresswell, Tim. 2012. "Mobilities II: Still." *Progress in Human Geography* 36 (5): 645–653. doi: 10.1177/0309132511423349.

Cresswell, Tim. 2014. "Mobilities II: Moving On." *Progress in Human Geography* 38 (5): 712–721. doi: 10.1177/0309132514530316.

Damonte, Gerardo H. 2018. "Mining Formalization at the Margins of the State: Small-Scale Miners and State Governance in the Peruvian Amazon." *Development and Change* 49 (5): 1314–1335. doi: 10.1111/dech.12414.

Dell, Melissa. 2010. "The Persistent Effects of Peru's Mining *Mita*." *Econometrica* 78 (6): 1863–1903. doi: 10.3982/ECTA8121.

Dessertine, Anna. 2016. "From Pickaxes to Metal Detectors: Gold Mining Mobility and Space in Upper Guinea, Guinea Conakry." *The Extractive Industries and Society* 3 (2): 435–441. doi: 10.1016/j.exis.2016.02.010.

Dougherty, Michael L. 2015. "By the Gun or the Bribe: Firm Size, Environmental Governance and Corruption Among Mining Companies in Guatemala." Chr. Michelsen Institute, U4 Issue Paper, Number 17. https://open.cmi.no/cmi-xmlui/handle/11250/2475268.

Elliott, Anthony, and John Urry. 2010. *Mobile Lives*. New York: Routledge.

Fafchamps, Marcel, Michael Koelle, and Forhad Shilpi. 2017. "Gold Mining and Proto-Urbanization: Recent Evidence from Ghana." *Journal of Economic Geography* 17 (5): 975–1008. doi: 10.1093/jeg/lbw015.

Ferguson, James. 1999. *Expectations of Modernity: Myths and Meanings of Urban Life on the Zambian Copperbelt*. Berkeley: University of California Press.

Frederiksen, Tomas, and Matthew Himley. 2020. "Tactics of Dispossession: Access, Power, and Subjectivity at the Extractive Frontier." *Transactions of the Institute of British Geographers* 45 (1): 50–64. doi: 10.1111/tran.12329.

Gadgil, Madhav, and Ramachandra Guha. 1992. *This Fissured Land: An Ecological History of India*. Berkeley: University of California Press.

Galeano, Eduardo. 1971. *Las venas abiertas de América Latina*. México, DF: Siglo XXI Editores.

Glave, Miguel. 1989. *Trajinantes: Caminos indígenas en la sociedad colonial siglos XVI-XVII*. Lima: Instituto de Apoyo Agrario.

Harvey, David. 1996. *Justice, Nature & the Geography of Difference*. Malden, MA: Blackwell Publishing.

Himley, Matthew. 2013. "Regularizing Extraction in Andean Peru: Mining and Social Mobilization in an Age of Corporate Social Responsibility." *Antipode* 45 (2): 394–416. doi: 10.1111/j.1467-8330.2012.01001.x.

Himley, Matthew. 2016. "Mining, Property, and the Reordering of Socionatural Relations in Peru." In *Mining in Latin America: Critical Approaches to the New Extraction*, edited by Kalowatie Deonandan and Michael L. Dougherty, 208–226. New York: Routledge.

Hinojosa, Leonith. 2013. "Change in Rural Livelihoods in the Andes: Do Extractive Industries Make Any Difference?" *Community Development Journal* 48 (3): 421–436. doi: 10.1093/cdj/bst023

Isenberg, Andrew C. 2005. *Mining California: An Ecological History*. New York: Hill and Wang.

Jønsson, Jesper Bosse, and Deborah Fahy Bryceson. 2009. "Rushing for Gold: Mobility and Small-Scale Mining in East Africa." *Development and Change* 40 (2): 249–279. doi: 10.1111/j.1467-7660.2009.01514.x.

Jønsson, Jesper Bosse, and Deborah Fahy Bryceson. 2017. "Beyond the Artisanal Mining Site: Migration, Housing Capital Accumulation and Indirect Urbanization in East Africa." *Journal of Eastern African Studies* 11 (1): 3–23. doi: 10.1080/17531055.2017.1287245.

Kemp, Deanna, Carol J. Bond, Daniel M. Franks, and Claire Cote. 2010. "Mining, Water and Human Rights: Making the Connection." *Journal of Cleaner Production* 18 (15): 1553–1562. doi: 10.1016/j.jclepro.2010.06.008.

Lane, Kris. 2019. *Potosí: The Silver City That Changed the World*. Oakland: University of California Press.

Larmer, Miles. 2021. *Living for the City: Social Change and Knowledge Production in the Central African Copperbelt*. Cambridge: Cambridge University Press.

Libassi, Matthew. 2022. "Gold Conflict and Contested Conduct: Large- and Small-Scale Mining Subjectivities in Indonesia." *Geoforum*. doi: 10.1016/j.geoforum.2022.10.005.

Loayza, Norman, and Jamele Rigolini. 2016. "The Local Impact of Mining on Poverty and Inequality: Evidence from the Commodity Boom in Peru." *World Development* 84: 219–234. doi: 10.1016/j.worlddev.2016.03.005.

Macintyre, Martha. 2018. "Afterword – Places, Migration and Sustainability: Anthropological Reflections on Mining and Movement." *Sustainable Development* 26 (5): 501–505. doi: 10.1002/sd.1895.

Massey, Doreen. 1991. "A Global Sense of Place." *Marxism Today* 35 (6): 24–29.

Moodie, T. Dunbar, and Vivienne Ndatshe. 1994. *Going for Gold: Men, Mines, and Migration*. Berkeley: University of California Press.

Moore, Jason W. 2007. "Silver, Ecology, and the Origins of the Modern World, 1450-1640." In *Rethinking Environmental History: World-System History and Global Environmental Change*, edited by Alf Hornborg, J. R. McNeill, and Joan Martinez-Alier, 123–142. New York: AltaMira Press.

Owen, John R., and Deanna Kemp. 2015. "Mining-Induced Displacement and Resettlement: A Critical Appraisal." *Journal of Cleaner Production* 87: 478–488. doi: 10.1016/j.jclepro.2014.09.087.

Perreault, Tom. 2013. "Dispossession by Accumulation? Mining, Water and the Nature of Enclosure on the Bolivian Altiplano." *Antipode* 45 (5): 1050–1069. doi: 10.1111/anti.12005.

Rivera Cusicanqui, Silvia. 2018. *Un mundo ch'ixi es posible: Ensayos desde un presente en crisis*. Buenos Aires: Tinta Limón.

Robins, Nicholas A. 2011. *Mercury, Mining, and Empire: The Human and Ecological Cost of Colonial Silver Mining in the Andes*. Bloomington: Indiana University Press.

Rohrbough, Malcolm. 2000. "No Boy's Play: Migration and Settlement in Early Gold Rush California." *California History* 79 (2): 25–43. doi: 10.2307/25463687.

Scott, Heidi V. 2009. *Contested Territory: Mapping Peru in the Sixteenth and Seventeenth Centuries*. Notre Dame, IN: University of Notre Dame Press.

Shaler, Andrew. 2020. "Indigenous Peoples and the California Gold Rush: Labour, Violence and Contention in the Formation of a Settler Colonial State." *Postcolonial Studies* 23 (1): 79–98. doi: 10.1080/13688790.2020.1725221.

Sheller, Mimi. 2018. "Theorising Mobility Justice." *Tempo Social, Revista de Sociologia de USP* 30 (2): 17–34. doi: 10.11606/0103-2070.ts.2018.142763.

Sheller, Mimi, and John Urry. 2006. "The New Mobilities Paradigm." *Environment and Planning A* 38 (2): 207–226. doi: 10.1068/a37268.

Stavig, Ward. 2000. "Continuing the Bleeding of These Pueblos Will Shortly Make Them Cadavers: The Potosi Mita, Cultural Identity, and Communal Survival in Colonial Peru." *The Americas* 56 (4): 529–562. doi: 10.1017/S0003161500029837.

Storey, Keith. 2010. "Fly-In/Fly-Out: Implications for Community Sustainability." *Sustainability* 2 (5): 1161–1181. doi:10.3390/su205116.

Szablowski, David. 2002. "Mining, Displacement and the World Bank: A Case Analysis of Compania Minera Antamina's Operations in Peru." *Journal of Business Ethics* 39 (3): 247–273. doi: 10.1023/A:1016554512521.

Taniguchi, Nancy J. 2000. "Weaving a Different World: Women and the California Gold Rush." *California History* 79 (2): 141–168. doi: 10.2307/25463691.

Ticci, Elisa, and Javier Escobal. 2014. "Extractive Industries and Local Development in the Peruvian Highlands." *Environment and Development Economics* 20 (1): 101–126. doi: 10.1017/S1355770X13000685.

Udelsmann Rodrigues, Cristina, Patience Mususa, Karen Büscher, and Jeroen Cuvelier. 2021. "Boomtown Urbanization and Rural-Urban Transformation in Mining and Conflict Regions in Angola, the DRC and Zambia." *Sustainability* 13 (4): 2285. doi: 10.3390/su13042285.

Uparela, Paola. 2022. "'Multiplicarse ha la tierra de gente y de fruto': Gender and Re-Production in Las Casas's and Guaman Poma's Biopolitical Projects (1516, 1615)." In *The Transatlantic Las Casas*, edited by David T. Orique and Rady Roldán-Figueroa, 35–60. doi: 10.1163/9789004515918_004.

Urry, John. 2007. *Mobilities*. Cambridge: Polity Press.

Verweijen, Judith, Sara Geenen, and Anuarite Bashizi. 2022. "Articulating Sedimented Subjectivities: Extractive Subject Formation in Eastern DRC." *Geoforum*. doi: 10.1016/j.geoforum.2022.10.009.

Yusoff, Kathryn. 2019. *A Billion Black Anthropocenes or None*. Minneapolis: University of Minnesota Press.

Zhu, Annah L., and Brian Klein. 2022. "The Rise of Flexible Extraction: Boom-Chasing and Subject-Making in Northern Madagascar." *Geoforum*. doi: 10.1016/j.geoforum.2022.06.005.

Zhu, Annah, and Nancy L. Peluso. 2021. "From Gold to Rosewood: Agrarian Change, High-Value Resources, and the Flexible Frontier-Makers of the Twenty-First Century." In *The Routledge Handbook of Critical Resource Geography*, edited by Matthew Himley, Elizabeth Havice, and Gabriela Valdivia, 345–357. New York: Routledge.

Section I

The Andes

2 *Ch'ixi* mobilities

Small-scale mining and Indigenous autonomy in the Bolivian tin belt

Andrea Marston

Introduction

In 2016, I conducted a series of life history interviews with small-scale tin miners in the towns of Llallagua and Uncía, which are located on opposite sides of Juan del Valle Mountain in the province of Rafael Bustillo, in the region of Bolivia known as Norte Potosí. One of these interviews was with Carlos, a garrulous miner in his early 30s. Like many other interlocutors, Carlos moved to Uncía in the 1990s to finish high school after growing up in the *ayllu* Kharacha, which surrounds and incorporates the municipality of Uncía (see Figure 2.1). An *ayllu* is a territorialized political formation common among Quechua and Aymara people in the Andes; although he now calls Uncía home, Carlos continues to make weekly returns to Kharacha to bring much-needed supplies and to assist his extended family with activities such as planting, harvesting, and freeze-drying potatoes into *chuño*. The rest of his time is split between studying law at Uncía's local university and mining for tin in the Juan del Valle Mountain. In other words, over the course of a week, Carlos moves between rural fields, urban classrooms, and subterranean mineshafts in a three-dimensional circuit that articulates multiple economic strategies as well as multiple places.

Carlos belongs to a mining cooperative, or an association of small-scale miners that holds a contract to exploit subterranean resources within a given area. Mining cooperatives are legal in Bolivia, but they are not entirely formal, in that most fail to comply with all relevant regulations. At present, there are seven mining cooperatives, which together incorporate more than 2,500 miners, operating in and around the Juan del Valle Mountain in Norte Potosí. The miners come from a range of backgrounds, including many who used to work for the state mining corporation, COMIBOL (*Corporación Minera de Bolivia*), before it was gutted by neoliberal austerity policies in 1985. But many other cooperative miners, like Carlos, have entered the mining business more recently, after migrating from one of the Norte Potosino *ayllus*. Norte Potosí is known for the strength of its *ayllus* (Mendoza Torrico and Patzi González 1997), but, for reasons discussed in this chapter, relatively few people migrated from the *ayllus* to the tin mining towns prior to the industry's neoliberalization in the late 1980s. Since then, however, *ayllu* members have become as numerous within the tin mining cooperatives as their ex-salaried counterparts. While forms of circular migration based on family and kinship connections

DOI: 10.4324/9781003313236-3

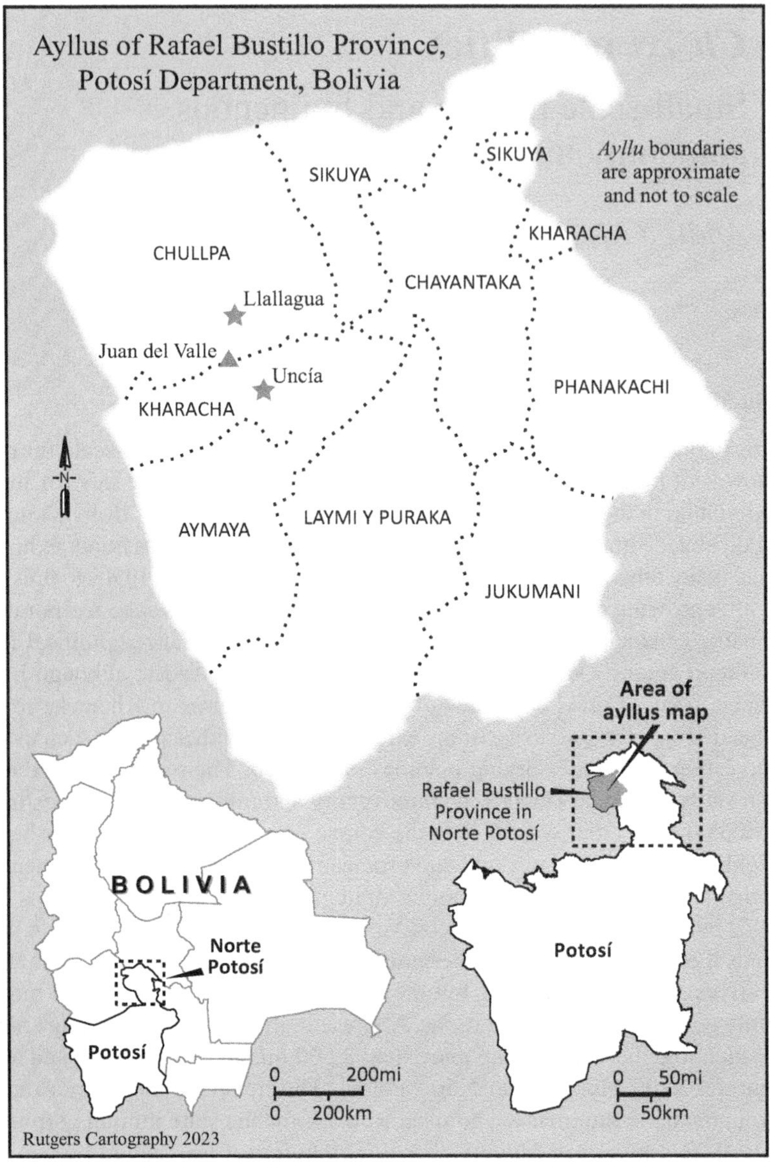

Figure 2.1 Map of the *ayllus* of Rafael Bustillo Province, in the Bolivian department of Potosí. Adapted from Mendoza Torrico and Patzi González (1997) by Rutgers Cartography. Note that *ayllus* are more porous geographical formations than they appear in most cartographic representations (Rivera Cusicanqui 2015), and this map is intended to give a sense of relative distance rather than indicate immutable spatial relations.

have long been observed around Andean mines (Harris and Albó 1976; Long and Roberts 1984), in Norte Potosí these mobility patterns now include more concerted involvement by Indigenous *ayllus* than in decades past.

What do these patterns mean for local Indigenous autonomy-building efforts? The struggle for Indigenous autonomy in Bolivia – which shares many concerns with demands for Indigenous sovereignty in North America (González 2015; Postero and Fabricant 2019) – has been gaining traction since at least the 1970s.[1] Although legal mechanisms to obtain state-recognized Indigenous autonomy have existed for at least a decade, this is not necessarily the only goal of the *autonomía* movement; state recognition is but one avenue toward political and economic self-governance (Albó and Barrios Suvelza 2006; Tockman and Cameron 2014). The economic piece of that puzzle, however, has often posed a significant hurdle for many Indigenous communities and *ayllus*. In the 1990s, NGOs and other nongovernmental entities assisted in the creation of important Indigenous federations, such as the highland federation CONAMAQ (*Consejo Nacional de Ayllus y Markas del Qullasuyu*), and in the early 2000s, the progressive political party MAS (*Movimiento al Socialismo*) – which held power from 2006 to 2019 and again since 2020 – offered some financial support. The current moment, however, is characterized by a lack of both state and non-state funding for strengthening Indigenous autonomy (Burman 2015). Coupled with the droughts that have been particularly challenging for agricultural communities in the Bolivian highlands, not to mention national-level economic shocks caused by the declining natural gas sector, this predicament has forced both individual *ayllu* members, as well as Indigenous leaders at *ayllu* and supra-*ayllu* levels, to look to mining as an economic activity with stabilizing potential.

This chapter suggests that a mobilities approach helpfully exposes some of the complex interconnections between Indigenous autonomy and small-scale mining in contemporary Norte Potosí, as well as demonstrating how indigeneity more broadly is constituted through movement. Rather than analyzing movement from one place to another on an otherwise static plane, mobilities research underscores how social worlds – and apparently distinct places and territories – are "emergent from complex and multi-scalar mobile relations, flows, circulations, and their temporary moorings" (Sheller 2018, 19). Attentive not only to physical movements but also to the meanings associated with those movements and to their embodied practices, this framework emphasizes how social formations are continuously reconstituted by people, things, and ideas "on the move." It is thus a powerful counterapproach to methodological sedentarism, in which territorial boundedness and static rootedness are presumed to be the norm, and migration is conceived as a temporary condition (Cresswell 2012). While bodies in general are often marked as different through specific movements and proximities (Sheller and Urry 2006), Indigenous people have been particularly subjected to methodological sedentarism – even definitionally contrasted against those who move, claim, and then settle (Li 2002). Approaching contemporary expressions of indigeneity within a mobilities framework can thus offer fresh insights into how Indigenous political fortification can occur through movement as well as place-making.

With this in mind, this chapter advances two arguments. First, small-scale tin mining in Norte Potosí is increasingly characterized by two relatively new forms of mobility: the movement of people between *ayllus* and nearby tin mines, and the regional circulation of mining-derived profits within rural areas. While mining in Bolivia has long relied on the forced migration and labor of Indigenous people, Indigenous communities have historically been excluded from the wealth derived from their extractive labors; now, money derived from mining circulates along with Indigenous miners themselves.[2] Second, these mobilities are thoroughly *ch'ixi* (motley) in that they involve the strategic use of capitalist economic activities to support contemporary expressions of Indigenous autonomy. As developed by Rivera Cusicanqui (2018) and her colleagues at Colectivx Ch'ixi, the Aymara word *ch'ixi* implies a coexistence of Indigenous and non-Indigenous worlds and practices that – unlike the twentieth-century ideology of *mestizaje* (miscegenation) – bends toward indigeneity. As used in this chapter, the concept supports a critical analysis of the degree to which new mining mobilities have the capacity to support Indigenous autonomy efforts.

I begin with a discussion of Andean mobilities in the precolonial and colonial eras before tracing the rise and fall of twentieth-century industrial tin mining. After exploring the role of Indigenous communities in the cooperative mining sector starting in the 1980s, I then examine more recent conversations about Indigenous autonomy and "communitarian mining." Finally, I conclude by reflecting on the possibilities and limits of *ch'ixi* mobilities, in which Indigenous autonomy is both secured and potentially undermined by twenty-first-century mining activities. Overall, this chapter suggests the mobilities approach's emphasis on the meanings and agencies associated with movement is what makes it particularly alive to these ambivalent experiences of small-scale mining.[3]

Mobilities and mining in Bolivia

People living in the Andes have long relied on a variety of mobility strategies to address the challenges associated with the mountainous terrain (Castillo and Brereton 2018). Pre-Incan political-economic formations revolved around the distribution of goods across multiple elevations as each *ayllu* sought "to control a maximum of floors and ecological niches" (Murra [1976] 1985, 16).[4] Ritually affirmed kinship relationships both lubricated the circulation of goods across this "vertical archipelago" and maintained the dominance of an *ayllu*'s highland administrative center vis-à-vis lowland "islands" that produced fruit and maize (Platt 2009). Andean communities were thus established *through* the movements of people and goods, and through the meanings attributed to those movements.

These vertical archipelagos changed with the consolidation of the Incan Empire in the late 1400s, and they were transformed further still with the consolidation of Spanish rule after 1532, but mobility remained a defining feature of both eras. The Incans ruled indirectly over pre-existing *ayllus* by positioning themselves as distant kin, as well as by imposing tributary taxation and labor drafting systems, the latter of which was known as the *mit'a,* or "turn" (Silverblatt 1987). All this depended on the continued mobilization of people and goods across elevations, now centralized to an unprecedented degree. When the Spanish took control of both taxation and *mit'a* systems, the latter went from incorporating a range of labors (including but not limited to mining) to being almost exclusively associated with labor in the Potosí silver mines. For the Spanish, the highlands were for mining and the valleys were for plantation-style agriculture, which meant that Aymara- and Quechua-speaking *ayllus* in the highlands were forced to pay tribute and send *mitayos* (drafted laborers) to the silver mines, while Quechua *ayllus* in the valleys were dismantled and their members were forced to work as *colonos* (sharecroppers) on the *haciendas* (agricultural estates) of landed elites (Gotkowitz 2007). The Spanish mining *mita* was expanded further still between 1574 and 1578, following Viceroy Toledo's reorganization of Indigenous communities: instead of being dispersed throughout the highlands, Andean people were forcibly concentrated into *reducciones*, settlements where their labor could be more easily taxed. This system sought to preserve the Incan system of vertical control even while significantly reorganizing social space (Mumford 2012).

Ongoing forced mobility permitted Spanish plunder under Toledo's regime, as one-seventh of all adult Indigenous men within the high-elevation *mita* catchment area were sent to the Potosí silver mines each year. Although they were technically supposed to work for one week at a time and then have two weeks off, in practice the need for income in the mining town forced *mitayos* to also become "free" workers, known as *mingas,* for the other two weeks as well. Even on the weekends, when they could not work as *mingas,* the miners went underground as *k'ajchas,* or subterranean thieves whose political power grew substantially over the centuries (Tandeter 1992; Abercrombie 1996). All this work was further supplemented by the labor of wives and children, who accompanied *mitayos* to Potosí (Barragán Romano 2020). After being resettled in *reducciones,* it took *mitayos* and their families – some 40,000 people total – at least two months to travel over mountainous terrain from Norte Potosí to the silver mines of the city, carrying food and bedding on their backs (Rowe 1957).

In (unequal) exchange for this forced mobility, some Bolivian *ayllus* were granted a degree of political autonomy within the Spanish empire, an arrangement that Tristan Platt (1982) calls the "pact of reciprocity." This autonomy was so important to the *ayllus* of Norte Potosí that they successfully rejected two attempts to subdivide their communal land into individual plots: first in 1874, when a liberal government tried to revitalize the economy by creating a "free" land market in Norte Potosí, and second in 1953, when a center-left government tried to apply a land reform that, while hugely popular in regions where plantation-style *haciendas* might be subdivided among landless workers, was hugely unpopular in regions that had maintained a collective hold over the land, as in Norte Potosí. Although the pact of reciprocity had been forged under colonial conditions, it remained preferable to the full-scale community dismantlement promised by land markets.

Nevertheless, *ayllus'* internal operations were transformed by the colonial economy with which they had been articulated. With the arrival of the colonial silver mining economy, Indigenous "kinship and alliance networks were rearticulated around regional and inter-regional markets, in an ambivalent deal with that new god, money, which not for nothing was associated with the figure of the devil"

(Rivera Cusicanqui 2018, 65).[5] This does not mean that money-as-universal-equivalent entirely displaced other forms of exchange; for instance, Olivia Harris (1989) writes about multiple kinds of money in the *ayllu* Laymi (see Figure 2.1), some of which could be acquired through exchanges with humans, some of which demanded exchanges with *wak'as* (earth beings), and all of which generated fertility and wealth through processes of circulation. Even when money-based forms of exchange may appear dominant, "more-than-capitalist" meanings are also part of this circulatory process.

Rivera Cusicanqui uses the Aymara word *ch'ixi* to describe the mixture of pre-colonial and colonial practices that persist in the present. The word literally describes a shade of gray that is produced as the effect of intermingling many black and white dots, as if by pointillism; rather than a synthesis or newly "pure" mixture, the two original colors persist alongside one another. Her larger interest in the concept relates to her dissatisfaction with both the ideology of *mestizaje*, which claims to celebrate racial-cultural mixture as perfect synthesis but actually privileges whiteness over indigeneity, and *indianismo*, which only valorizes a "pure" and romanticized form of indigeneity. She describes her sense of *ch'ixi* as a "decolonizing *mestizaje*," which exists outside the central state and celebrates internal contradictions (Rivera Cusicanqui 2018, 145). A central feature of this decolonizing project is a rethinking of the economy, and one long quote usefully summarizes her take on the market:

> The demonization of the market, is it not perhaps a way of simplifying, in search of the idyllic harmony of the "natural economy," another variant of the European dream of the noble savage? Is it not a way of denying that in this story the complexities can play in our favor? [...] I think that in popular languages and practices there is a margin of maneuver that makes it possible to manage multiple ecological niches, play various sides, various cards, but without losing one's direction [*sin perder el eje*] (70, translation by author).

In other words, the market is part of her *mundo ch'ixi* (motley world). Even if it has become hegemonically associated with capitalist accumulation, it is also part of a longer Andean tradition of exchange and can be subordinated to communal, rather than strictly individualized, interests. It is with Rivera Cusicanqui's intervention in mind that I am considering contemporary Andean engagements with mining, and the forms of mobility that underpin them, as potentially *ch'ixi*.

In my analysis, *ch'ixi* mobilities are the movements of people, things, or ideas that result not only in the articulation of Indigenous and non-Indigenous worlds (polities, economies, etc.) but also in the meaningful repurposing of the latter to defend the former; these mobilities bend toward indigeneity, not assimilation. The concept emphasizes the agential qualities of these movements, honoring the individual and collective decisions involved in deciding when, where, and how best to be on the move. In Norte Potosí, such decisions and their corresponding mobilities are inescapably intertwined with tin mining.

From enclave economies to mining cooperatives

About halfway up the Juan del Valle Mountain, the urban area of Llallagua gives way to a suburban zone that miners use to concentrate tin, a process that involves crushing ore rocks and using various configurations of water and gravity to capture the fine grains of tin. The concentration zone is lined with small square storage rooms, pools of water, and cement patios topped with *quimbaletes*, semi-circular rollers made of scrap metal and stones that miners use to transform ore rocks into sand (Figure 2.2).

Figure 2.2 A *quimbalete* in action. Uncía, 28 February 2016. Photo by author.

Miners usually concentrate ore on Saturdays, and in 2016–2017, I made a habit of joining them. On one such visit, I met a father and son named Wilfredo and José, respectively. Wilfredo was sifting through a barrel of ore slurry, using a sieve to remove the largest rock shards, while José was gently swirling ore sands in a large tray of water; the heavier tin particulates moved to the lower end of the tray, while the lighter waste rocks remained near the top. As they worked, they told me a bit about their trajectories into mining. Wilfredo grew up in rural Aymaya, an *ayllu* to the south of Llallagua (see Figure 2.1), and moved to town with his own father in 1987 to join a mining cooperative in its infancy. He has been working with the cooperative ever since, though he still returns regularly to Aymaya to care for his family's fields and livestock. José, by contrast, was born in Llallagua and had only been working in the mines for four years. At the time that we spoke, he was only in his early twenties, but he had already moved around a lot: he went to the city of Cochabamba at age 15 to work in construction, then he moved to the lowland department of Santa Cruz to work in logging, and went on to work as a waiter in São Paulo before returning to Llallagua to assist his father in the mine. He was considering traveling to Chile to work the grape harvest, and then perhaps staying on for tomatoes or olives, but he intended to return to Llallagua eventually to rejoin his father both in the mine and in their home *ayllu*, since the burden of both responsibilities was too much for the older man alone.

The details of Wilfredo's and José's stories are, of course, unique to them, but they nevertheless speak to certain trends within both Norte Potosí's *ayllus* and tin mining in Llallagua-Uncía. Since the late 1980s, an increasing number of people from the *ayllus* have started to mine under the legal auspices of mining cooperatives, though this is also happening in conjunction with a variety of other economic strategies (waged labor, informal/illicit labor, national and international migration, etc.). To understand why this is only a relatively recent phenomenon, however, one must return to the early twentieth century and to the sudden frenzied burst of tin mining that definitively changed the Norte Potosino landscape.

Although tin is an ancient metal, it had acquired a new economic importance in Europe in the late 1800s: in addition to a growing set of uses in manufacturing and armaments production, tin was suddenly celebrated for its ability to preserve food. Europe's traditional source of the metal, Cornwall, was struggling to keep pace, and prospectors around the world started to hunt for tin (Ingulstad, Perchard, and Storli 2014). Already in the 1890s, several British and Chilean companies had opened small mines in the places that today are called Llallagua and Uncía, but the Bolivian tin belt had not yet proven itself especially lucrative. It was a discovery by Simón I. Patiño, a Bolivian prospector from Cochabamba, that catapulted Norte Potosí into tin mining fame. Patiño's lucky strike in Llallagua allowed him to begin a quiet acquisition of all other mining interests in the region (Querejazú Calvo [1977] 1998). This move eventually turned him into one of three Bolivian "tin barons" – who collectively controlled 80.9 percent of all Bolivian tin mines by 1940 – as well as one of the wealthiest people in the world at the time of his death in 1947 (Granados 2014).

At the height of the tin business, Patiño's holdings in Llallagua and Uncía employed thousands of workers, not only in the mines but also in the *maestranza* (machine workshop), the *ingenios* (processing plants), the geology and engineering offices, and the *pulpería* (company store), among others. Very few of these workers, however, were from the *ayllus* of Norte Potosí. Historian Luis Oporto Ordóñez's (2007) study, which focuses on the early era of tin mining in Llallagua-Uncía (1900–1935), argues that workers in Uncía came largely from the exhausted silver mines of Colquechaca, within Norte Potosí, while workers in Llallagua came largely from the valleys of Cochabamba and Oruro. He did not find a single case of a worker born in Uncía prior to 1913. My own review of employee files in the COMIBOL archives shows similar results: of 300 workers hired between 1910 and 1929, 110 (37 percent) came from the department of Cochabamba, 55 (18 percent) were from the department of Oruro, 46 (15 percent) were from the department of La Paz, and 23 (eight percent) were from other parts of the department of Potosí. Although 41 workers were from Norte Potosí, nearly half of them were from Colquechaca, and only 22 (seven percent) were from villages and *ayllus*.[6] While it is impossible to know exactly why this was the case, one might hypothesize a mixture of factors, ranging from desire among Cochabamba *colonos* to leave *haciendas* to resistance within the communities of Norte Potosí, whose members preferred relative political autonomy over economic incorporation.

Regardless of why it happened, however, the tin mining industry emerged as something of an economic bubble within *ayllu* territory. Llallagua and Uncía were classic company towns in that workers relied on the *pulpería* for staples such as bread and meat, lived in company housing, made use of company-provided medical services, and sent their children to the company-run school. Although they were never *completely* isolated from the surrounding rural areas – for instance, miners' wives regularly traveled to rural markets to trade company store goods for local products (Godoy 1985a) – miners nevertheless tended to develop stronger ties with distant mining zones, with which they were connected via railroad, than they did with nearby communities. These salaried tin miners were a major driving force behind Bolivia's 1952 National Revolution, which resulted in the nationalization of all the mines held by the three tin barons and the creation of the state mining corporation (COMIBOL). All this upheaval, however, had relatively little impact on the invisible membrane that still seemed to separate the mines from the surrounding *ayllus*.

There is some danger of overstatement here, particularly in the years after the 1952 revolution. As suggested by the archival evidence, *some* people did come from the *ayllus*, even if they made up a small minority of the workforce, and there were more openings for them in the postrevolutionary decades. By the 1980s, several anthropologists noted that the full-time salaried workers inside the mining company were something of a "working class aristocracy" within the larger social pyramid of Llallagua, underneath which there was a diverse range of non-salaried workers, including members of "subsidiary organizations" (groups of miners who paid the company for the right to exploit less profitable corners of the mine), *palliris* (women who traversed slag heaps looking for discarded ore-bearing rocks), and *jukus* (ore thieves who operated at night) (Harris and Albó 1976; Godoy

1985b). But even if these non-salaried workers were growing in numbers, they did not constitute the voice or collective face of the tin miners, whose salaried and unionized members continued to dominate national political stages throughout the 1960s and 1970s.

As they struggled against dictatorships and imperialism, unionized tin miners also became representatives of the ideology of *mestizaje*, which had been embraced by the center-left government that took power after the 1952 National Revolution. Like the Mexican state's embrace of the "cosmic race," or a perfect racial synthesis that represented a new nationalism, Bolivia's *mestizaje* imagined a "pure" mixture that ultimately erased contemporary Indigenous people by locating them as shared ancestors in a national past (Rivera Cusicanqui 2003; Sanjinés C. 2004). The postrevolutionary government institutionalized *mestizaje* by reclassifying all Bolivians in class-based rather than race-based terms: *indios* became *campesinos* (peasants), *criollos* (Spanish descendants) became professionals, and all were officially considered *mestizo*. Salaried tin miners – most of whom were Quechua-speaking descendants of Cochabamba *colonos* or intergenerational miners, and who remained relatively separated from the *ayllus* of Norte Potosí – already identified more with the global proletariat and an emergent popular nationalism than they did with indigeneity, in no small part because of the anti-Indigenous racism that permeated their schools and mandatory military trainings (Gill 1997). Unionized miners thus became emblems of a new, *mestizo* nationhood that, although it may have embraced more people than its republican counterpart, continued to deny or erase Indigenous autonomies.

The unions of Llallagua and Uncía, however, were largely dismantled starting in 1985, a moment that miners sometimes refer to as the "neoliberal crack." Crisis in the global tin market, coupled with a generalized debt crisis across Latin America, pushed the Bolivian government to accept foreign loans predicated on "de-nationalizing" the mining sector, among other things. Because there was very little international interest in investing in a failing industry, most of COMIBOL's tin mines were either much reduced in size or simply closed, while around 30,000 miners were laid off nationwide (Kohl and Farthing 2006).

This is the point in the story at which mining cooperatives become important. While there had been a growing number of mining cooperatives throughout the country since the General Law of Cooperative Societies (#5053) was passed in 1958, cooperatives became the solution of choice for the neoliberal government looking to contain politically unruly – and now unemployed – tin miners, who had started migrating to every corner of the country (Poveda Ávila 2014). The government was particularly concerned by an uptick in the number of ex-miners entering the coca-growing sector, so 1987 saw the introduction of another decree (DS21377) designed to encourage miners to move *back* to the mines: a decree permitting the formation of cooperatives within COMIBOL holdings.

Key to this chapter, however, is that mining cooperatives were not *only* composed of ex-salaried miners. Instead, a large number of people migrated from the rural areas of Norte Potosí to work in Llallagua and Uncía. This is the moment in which both Wilfredo, introduced in this section, and Carlos, introduced at the beginning of the chapter, began mining. Again, it is hard to say exactly why neoliberalism seemed to pierce the barrier between the enclaved tin industry and the surrounding *ayllus*, but my interlocutors suggested a combination of less racial discrimination (compared to hiring practices within either Patiño's company or COMIBOL) and more fluidity within the cooperative structure, which allowed cooperative members to come and go from the mining towns in ways that suited their familial and community obligations. Although *ayllu* members did not necessarily enter on the same economic footing as their ex-salaried counterparts – there was still racism, even if it no longer prevented entry into the mines – many of them stayed and continued into leadership roles, in which capacities they were able to encourage even more *ayllu* participation. Their individual mobilities thus encouraged more widespread mobility across the *ayllus* at a regional level.

This trend continued and deepened in the early 2000s, when mining cooperatives grew rapidly in a context of both high commodity prices and favorable political conditions. While mineral and metal prices both soared, left-leaning President Evo Morales, who came to power in 2006, focused his efforts on developing the natural gas sector rather than fully reactivating COMIBOL (Webber 2011). In part because cooperative miners had been active supporters during his presidential campaigns, Morales assisted the cooperative mining sector in a variety of ways, including appointing a cooperative miner as his first Minister of Mining and Metallurgy, using profits derived from the natural gas sector to "gift" cooperatives with machinery and equipment, and streamlining the bureaucratic processes required to form a cooperative. All of this encouraged the movement of people into the cooperative mining sector, and in Norte Potosí most of these "newcomers" were from the regional *ayllus*. In 2016–2017, more than half of the miners with whom I spoke had migrated to the mining towns relatively recently from surrounding *ayllus*, especially Chullpa (which contains the municipality of Llallagua) and Kharacha (which contains the municipality of Uncía – see Figure 2.1). Chullpa has been especially well-represented since some of its members occupied and blockaded the mine in 2005; this conflict was resolved with the formal admission of 100 Chullpa members into Llallagua's largest cooperative.

The relationship that these "new" miners maintain with their home *ayllus* is highly varied, but most people travel back to their hometowns at least several times per year to assist with planting and harvesting, and to participate in local festivals. When they return, they bring gifts of food, drink, supplies, and money, which ensures that at least part of the wealth derived from mining arrives in Indigenous communities. Others are even more strategic about their approach to the possible mobility of money, encouraging their home *ayllus* to tax mining members to create a kind of internal development fund. For instance, I spoke at length with a cooperative miner from Chullpa who was trying to convince his *ayllu*'s leadership to pool money from mining and put it toward building a brewery that could function as a community corporation. Whether this plan comes to pass remains unclear, but the vision is certainly there.

Contemporary cooperative miners tend to elicit a range of negative reactions from Bolivians, from those who imagine cooperative miners as thieves of national resource wealth (because they pay low royalties and no taxes) to those who associate them with a developmental "regression" – one interlocutor, a geologist at the local university in Llallagua, described cooperatives as "walking backwards like crabs." While there are a lot of concerns about cooperative mining from an environmental standpoint, some of the antipathy toward the sector is anti-Indigenous. For instance, a faculty member at Llallagua's local university, in the mining engineering department, described cooperatives thus:

> The problem with the cooperatives is that most people come from the countryside. They were never [politically] trained (*orientados*). Their leaders are very egotistical – they think they know more than an engineer. They don't let us enter. In vulgar terms, they are stingy (*mesquinos*). Those same leaders come from the countryside, and with this change of government [Morales's government] they have become more combative.
>
> (Interview, Llallagua, 2 February 2017)

Describing someone as "from the countryside" in Norte Potosí implies they are from the *ayllus*, meaning that the phrase's racial content is concealed in a geographic reference. In this case, the engineer linked origins in the countryside to a lack of political consciousness and also attributed cooperative miners' newly "combative" stances to the support they had received under Evo Morales's government. As did many other interviewees, he viewed this militancy in a negative light and seemed to imply that cooperative miners were not staying in their place – the countryside. They had become too mobile. While this sentiment echoes a more general distinction between the figure of the *indio permitido* (who adheres to prescribed "cultural" expectations while not interfering with capitalist development) and every other expression of indigeneity (Hale and Millamán 2006), this interview also specifically highlighted the role of (im)mobility in regional hierarchies of race and indigeneity.

The idea of the *indio permitido* was originally articulated by Silvia Rivera Cusicanqui in reference to the limitations of cultural rights (orally, in a moment of apparent exasperation – see Hale [2004] for discussion), and I imagine the *indio permitido* as the figure that a *ch'ixi* world is attempting to move beyond. Instead of expectations of purity – either of *mestizaje* or *indianismo* or even liberal multiculturalism – the notion of *ch'ixi* describes the non-synthetic coexistence of multiple tendencies, and even multiple expressions of indigeneity. This chapter suggests that contemporary cooperative miners who maintain connections with the *ayllus* of Norte Potosí do so through *ch'ixi* mobilities. Outside of Norte Potosí, a growing call for "communitarian mining" similarly rests on an interest in promoting these interconnections.

The call for communitarian mining

The launch of a new constitution in 2009 was a watershed moment for Bolivia, marking the country's transition from a "republic" into a "plurinational state" and enshrining its commitment to both Indigenous values (such as reciprocity and non-extractive relations with nature) and Indigenous autonomies. This constitution, however, also laid the groundwork for contemporary mining regimes. It recognizes three legal "actors" in the mining sector: the state corporation (COMIBOL), private companies, and mining cooperatives. Anyone who wishes to begin mining, regardless of whether they own the land above the known or suspected mineral deposit, must belong to one of these three entities. This has essentially functioned as a vertical territorial limit on Indigenous autonomies, given that subterranean spaces beneath collectively held land can be contracted out to other entities, while Indigenous communities cannot themselves begin mining without seeking permission from the (still colonial) state.

In recent years, however, several Indigenous communities and activists in Bolivia have called this vertical separation into question. In 2013, in the leadup to the approval of a new comprehensive mining law (passed in 2014), the highland Indigenous federation CONAMAQ organized a national summit to discuss the possibility of demanding formal recognition for what they called "communitarian mining." They hoped that communitarian mining could become a fourth actor in the mining sector, recognized by law if not by the constitution (Marston and Kennemore 2019). Although unsuccessful at the time, this demand was raised again, nearly a decade later, in the context of rising mineral prices. On 18 May 2022, President Luis Arce's government released a supreme decree that included an article calling for the development of "policies that support Indigenous participation in the benefits of the mining industry" (DS4721 Article 3: II: i). Although neither the mining law nor the constitution has been altered to welcome a fourth actor into the mining sector, this decree seems to suggest that such legal harmonization might be possible in the future.

In the interim between communitarian mining's initial proposal and its current rise to prominence, I interviewed one of the people who led the original charge. Tata Félix Becerra had been the leader of CONAMAQ from 2011 to 2013; at the time that we spoke in October 2016, however, he was imprisoned in San Pedro, La Paz's minimum-security prison for men, after having been accused of using money intended for development projects in his home *ayllu* for personal use. His case was part of a much larger scandal around the *Fondo de Desarrollo Para los Pueblos Indigenas Originarios y Comunidades Campesinas* (Fund for the Development of Original Indigenous Peoples and Campesino Communities, or FONDIOC) that had exploded in late 2015; Tata Félix was one of several Indigenous leaders to have been imprisoned after it was "revealed" that the vast majority of Fund-financed projects had never materialized. For Indigenous nations and their allies, this scandal and its accompanying incarcerations constituted an enormous set of smoke and mirrors designed to discredit Indigenous leaders who no longer supported the Morales government. Tata Félix's leadership period had coincided with CONAMAQ's move to distance itself from Evo following the 2011 TIPNIS conflict,[7] and animosities only deepened after he organized the communitarian mining summit in 2013. By the time I spoke with Tata Félix, CONAMAQ was virtually paralyzed for lack of funding, which is partly why he had been pushing for communitarian mining – as a way of funding the cash-strapped federation. In short, communitarian mining was the reason for my visit, but it was likely also one of the reasons why Tata Félix had been imprisoned in the first place, given that it was one of several ways he had fallen out with the MAS government.

I was not able to record this interview because of prison regulations against technology, so the quotes below are approximations derived from furious notebook scribbling:

> Look, the struggle for communitarian mining was always about the struggle for autonomy, about getting control over the non-renewable resources in our territories. We are still doing communitarian mining, even though it was never approved. We work within the *marco* (legal framework) of the mining cooperativas, but we are self-governing Indigenous communities. We practice our *usos y costumbres* [uses and customs] within the [cooperative] mining process. [...] Everyone is automatically affiliated to both the Indigenous federation and to the mining cooperative.
>
> (Interview, La Paz, 19 October 2016)

In other words, what his *ayllu* in Cochabamba was doing in practice was organizing a cooperative. But the difference between their cooperative and a typical cooperative, Tata Félix assured me, was in the details of their more egalitarian distribution of access rights and financial benefits. He explained that everyone who was performing their communal responsibilities was assigned a three-dimensional "plot" upon reaching adulthood. The only expense was a fixed amount paid annually (100Bs, or about $12.50) to cover the rental cost of the mining contract and some collective meetings. Everyone paid the same amount whether they were actively mining or not; that way the mine supported the *ayllu*. Indeed, the more he spoke, the more it seemed he was making an argument about Indigenous economic autonomy, though this was not the language he used. In his words:

> We produced coca, wheat, potatoes – but these have low prices. Slowly we decided we needed to administer our non-renewable resources. Indigenous communities are smart; we wanted to be owners. We knew our rights to manage our territory. We knew the laws, knew about contracts with international companies. In 2013, we wanted to move forward. That's when we had the [communitarian mining] summit. [...]
>
> All minerals belong to the state, it's the owner. It didn't pay attention to Indigenous communities, and in the constitutional assembly they just associated us with llamas, *chuño* [freeze-dried potatoes], potatoes – they assumed we would live from that. But that's not the case! We have rights to *all* our resources. In 2006 we were already proposing communitarian mining, but in reality, we don't care what it's called. What's important is the idea of Indigenous autonomy: we can administer exploitation. It's territory all the way down (*territorio hasta abajo*).
>
> (Interview, La Paz, 19 October 2016)

This political vision shreds any romantic notion that Indigenous peoples are necessarily opposed to non-renewable resource extraction. In fact, movement into subterranean spaces could facilitate the economic autonomy that is so often missing from political autonomy.

This aspiration did not die when Tata Félix left office. When I was in Llallagua-Uncía in 2016–2017, the leader of CONAMAQ happened to be living in Uncía, where his children were attending high school and college. Tata Max was from the *ayllu* Aymaya which, although it was just a little distance beyond Kharacha, did not seem to send as many people work in the cooperatives. Instead, he – like Tata Félix before him – was interested in seeing if his *ayllu* and others might start their own co-operatives, in the communitarian model suggested by Félix. Much later, in 2022, non-Indigenous interviewees in mining towns across the *altiplano* spoke nervously about the threat of *avasallamiento* (occupation/takeover) by rural Indigenous communities. Although there has been a law against *avasallamiento* since 2013 – a law that has been accused of criminalizing anti-mining protest – it has not prevented an increasing number of mine takeovers. Some of these *avasallamientos* aim at shutting down mining activities, but others are about claiming ownership over the extractive process.

In her discussion of *ch'ixi* landscapes, Penelope Anthias (2017) describes how the defense of Indigenous territorial rights in a context of a natural gas bonanza in the Bolivian Chaco region involves an engagement with multiple and competing land claims, values, and world views. In other words, a *ch'ixi* landscape is a spatialization of Rivera Cusicanqui's (2018) insistence on the existence of multiple value forms in *ch'ixi* economies. To this, I would add that these landscapes emerge through multiple forms of mobility and their associated meanings. In the case of Norte Potosí, people from *ayllus* move between mining towns and home communities but intend their labor in the former to be supportive of the latter. At a larger scale, Indigenous leaders recommend that a collective movement into subterranean spaces might be key to strengthening Indigenous autonomies. The mobilities implied between traditional and not-so-traditional spaces are thoroughly *ch'ixi,* or mixed in a way that is intended to strengthen (rather than erase or dilute) Indigenous governance. These are agential decisions, made under challenging economic circumstances but made nonetheless, and the mobilities approach allows them to be interpreted as such.

Conclusion

Mobility in the service of mining has been part of life in the *ayllus* since the Incan empire introduced the *mit'a* obligation, though it was the Spanish empire that turned this tributary system into a structuring component of *ayllu* autonomy: only by sending *mitayos* and their families to the Potosí silver mines could the *ayllus* of Norte Potosí safeguard their own relative independence from the colonial state. In other words, Norte Potosinos were mobile miners out of necessity, and their participation in silver mining dropped significantly after the *mita* system was terminated in 1812. The tin mining industry that took root within Norte Potosí at the turn of the twentieth century largely excluded local *ayllu* members from its workforce, and it was not until after the collapse of the state mining company in 1985 that *ayllu* members began both mining within their own territories and profiting from the activity directly. This change has involved a variety of human mobilities, including circular migration between mines and *ayllus*; it has also involved changing patterns in the mobility of money, which now circulates regionally rather than exiting the zone *en route* to private banks and state treasuries.

This chapter should not be read as an unqualified celebration of mining as an avenue to Indigenous autonomy. The environmental and health-related implications of most mining activities are well-documented, including within the tin belt of Norte Potosí (López Canelas 2010; Perreault 2013). The turn to mining within the region's *ayllus* might be better described as the result of the double bind of contemporary iterations of colonialism, in which an engagement with colonial economies is necessary to stage a defense against colonial polities. In recent decades, financial support for Indigenous federations such as CONAMAQ – which uses the funds to host gatherings, maintain office spaces, and provide travel allowances for leaders – has come primarily from either pro-Indigenous NGOs or from supportive governments; in the absence of both, leaders are pushed to seek an alternative source. While neither NGO or government investment was free from contradictions, many activists and researchers find mining far more perplexing, perhaps because it cannot be easily squared with environmental stewardship. Yet particularly given the host of other challenges faced within *ayllus,* including droughts and soil pollution – not to mention the high prices fetched by metals on the global market – mining has become an increasingly attractive option.

The *ch'ixi* worlds and *ch'ixi* economies discussed by Rivera Cusicanqui, as well as the *ch'ixi* landscapes elaborated by Anthias, all point to the limitations of conversations that posit an impermeable separation between Indigenous and non-Indigenous worlds (whether imagined as *mestizo* or colonial or white). At stake in the multiple iterations of the concept of *ch'ixi* is a reimagining of Indigenous economies and territories as potentially strengthened through its members' mobilities. Indeed, it is through an examination of mobility that I have been considering the relationship between *ayllus* and mining in Norte Potosí, and it is through this lens that I have been able to think about how Indigenous autonomies might be strengthened through carefully chosen external connections, all of which are forged in movement. The distance that has historically separated mining towns and *ayllus,* or even the aboveground from the below-ground, has often been conceptual as much as geographical: most studies from the region either focus on mining or *ayllu* politics, rather than the connections between the two. To the extent that both topics *are* explored together, their relationship is often presumed to be not only external but also antagonistic. Focusing on the mobile activities that unite the two spaces, as this chapter has done, can reveal a much more complicated set of dynamics, in which people use movement to assert their commitments to community, autonomy, and place.

Notes

1 In the early 1970s, a group of Indigenous intellectuals near the city of La Paz began calling themselves *Kataristas*, taking their name from eighteenth-century Indigenous revolutionary Tupak Katari. They sought to understand the "double oppression" of Indigenous peasants under both colonialism and capitalism.

2 See Absi (2005) for a similar argument about cooperative miners and the circulation of money around the city of Potosí.
3 Methodologically, this chapter draws on a mixture of interviews, participant observation, and archival analysis undertaken in 2016, 2017, and 2022. I interviewed cooperative miners, highland Indigenous leaders, unionized miners, and a range of experts based in La Paz, Oruro, Cochabamba, Huanuni, and Llallagua-Uncía, and I worked primarily in the COMIBOL archives (in El Alto and Catavi) as well as in private archives based in La Paz and Cochabamba.
4 Although the complexity of *ayllu* organization is not the focus of this chapter, it is important to note that *ayllus* are not organized into neatly nested political territories like nations-provinces-municipalities. Instead, major *ayllus* are usually divided into upper and lower halves (Anansaya and Urinsaya, respectively), each half is divided into minor *ayllus,* and some *ayllus* are further divided into small units called *cabildos.* Each of these units may have its "own" settlements at other altitudes, and these settlements may be located geographically inside other *ayllus,* all of which has been perplexing to western cartographers (Harris and Velasco 1997).
5 The devil figure, or El Tío, controls subterranean veins and still receives offerings of coca leaves, cigarettes, and liquor in mines throughout Bolivia.
6 These files were randomly sampled. Of the remaining 25 employees, nine were from Chuquisaca, one each was from Santa Cruz and Tarija, six were from foreign countries (Argentina, Chile, Peru, and Yugoslavia), and the files of the rest were either illegible or blank. Results taken from the COMIBOL Archives in Catavi, a submunicipality of Llallagua.
7 In 2011, a conflict emerged after President Morales announced plans to build a highway from the tropics of Cochabamba to the Brazilian border, slated to pass directly through lowland Indigenous territories and a national park called TIPNIS (*Territorio Indígena y Parque Nacional Isiboro Sécure,* Isiboro Sécure Indigenous Territory and National Park). After that, Bolivia's two major Indigenous federations – CONAMAQ in the highlands and CIDOB (*Confederación de Pueblos Indígenas de Bolivia*, Confederation of Indigenous Peoples of Bolivia) in the lowlands – publicly denounced the gap between Morales's discourse and action (Burman 2014; Fabricant and Postero 2015).

References

Abercrombie, Thomas A. 1996. "Q'aqchas and la Plebe in 'Rebellion': Carnival vs. Lent in 18th-Century Potosí." *Journal of Latin American Anthropology* 2 (2): 62–111.

Absi, Pascale. 2005. *Los ministros del diablo: El trabajo y sus representaciones en las minas de Potosí.* La Paz: Fundación PIEB.

Albó C., Xavier, and Franz X. Barrios Suvelza. 2006. *Por una Bolivia plurinacional e intercultural con autonomías.* La Paz: IDH Bolivia.

Anthias, Penelope. 2017. "Ch'ixi Landscapes: Indigeneity and Capitalism in the Bolivian Chaco. *Geoforum* 82: 268–275.

Barragán Romano, Rossana. 2020. "Women in the Silver Mines of Potosí: Rethinking the History of 'Informality' and 'Precarity' (Sixteenth to Eighteenth Centuries)." *International Review of Social History* 65 (2): 289–314.

Burman, Anders. 2014. "'Now We Are Indígenas': Hegemony and Indigeneity in the Bolivian Andes." *Latin American and Caribbean Ethnic Studies* 9 (3): 247–271.

Burman, Anders. 2015. "El ayllu y el indianismo: Autenticidad, representatividad y territorio en el quehacer político del CONAMAQ, Bolivia." In *Los nuevos caminos de los movimientos sociales en Latinoamérica*, edited by A. M. Ejdesgaard Jeppsen, H. Balslev Clausen, and M. A. Velázquez García, 100–122. Monterrey, Mexico: Tilde Editores.

Castillo, Gerardo, and David Brereton. 2018. "Large-Scale Mining, Spatial Mobility, Place-Making and Development in the Peruvian Andes." *Sustainable Development* 26 (5): 461–470.

Cresswell, Tim. 2012. "Mobilities II: Still." *Progress in Human Geography* 36 (5): 645–653.

Fabricant, Nicole, and Nancy Postero. 2015. "Sacrificing Indigenous Bodies and Lands: The Political–Economic History of Lowland Bolivia in Light of the Recent TIPNIS Debate." *The Journal of Latin American and Caribbean Anthropology* 20 (3): 452–474.

Gill, Lesley. 1997. "Creating Citizens, Making Men: The Military and Masculinity in Bolivia." *Cultural Anthropology* 12 (4): 527–550.

Godoy, Ricardo A. 1985a. "State, Ayllu, and Ethnicity in Northern Potosí, Bolivia." *Anthropos* 80: 53–65.

Godoy, Ricardo A. 1985b. "Technical and Economic Efficiency of Peasant Miners in Bolivia." *Economic Development and Cultural Change* 34 (1): 103–120.

González, Miguel. 2015. "Indigenous Territorial Autonomy in Latin America: An Overview." *Latin American and Caribbean Ethnic Studies* 10 (1): 10–36.

Gotkowitz, Laura. 2007. *A Revolution for Our Rights: Indigenous Struggles for Land and Justice in Bolivia, 1880–1952.* Durham, NC: Duke University Press.

Granados, Oscar. 2014. "Bankers, Entrepreneurs, and Bolivian Tin in the International Economy, 1900-1932." In *Tin and Global Capitalism, 1850-2000: A History of "The Devil's Metal",* edited by Mats Ingulstad, Andrew Perchard, and Espen Storli, 46–73. New York: Routledge.

Hale, Charles. 2004. "Rethinking Indigenous Politics in the Era of the 'Indio Permitido'." *NACLA Report on the Americas* 38 (2): 16–21.

Hale, Charles R., and Rosamel Millamán. 2006. "Cultural Agency and Political Struggle in the Era of the Indio Permitido." In *Cultural Agency in the Americas*, edited by Doris Sommer, 281–304. Durham, NC: Duke University Press.

Harris, Olivia. 1989. "The Earth and the State: The Sources and Meanings of Money in Northern Potosi, Bolivia." In *Money and the Morality of Exchange*, edited by Jonathan Parry and Maurice Bloch, 232–268. Cambridge: Cambridge University Press.

Harris, Olivia, and Xavier Albó. 1976. *Monteras y guardatojos. Campesinos y mineros en el Norte de Potosí.* La Paz: CIPCA.

Harris, Olivia, and Luisa Fernanda Velasco. 1997. *Somos los hijos de los ayllus: Pasado y presente de los pueblos indígenas del Norte de Potosí.* La Paz: Ministerio de Desarrollo Humano.

Ingulstad, Mats, Andrew Perchard, and Espen Storli, eds. 2014. *Tin and Global Capitalism, 1850-2000: A History of "the Devil's Metal"*. New York: Routledge.

Kohl, Benjamin, and Linda Farthing. 2006. *Impasse in Bolivia: Neoliberal Hegemony and Popular Resistance.* London: Zed Books.

Li, Tania Murray. 2002. "Ethnic Cleansing, Recursive Knowledge, and the Dilemmas of Sedentarism." *International Social Science Journal* 54 (173): 361–371.

Long, Norman, and Bryan Roberts. 1984. *Miners, Peasants and Entrepreneurs: Regional Development in the Central Highlands of Peru.* Cambridge: Cambridge University Press.

López Canelas, Elizabeth, ed. 2010. *El costo ecológico de la política minera en Huanuni y Bolívar.* La Paz: Fundación PIEB.

Marston, Andrea, and Amy Kennemore. 2019. "Extraction, Revolution, Plurinationalism: Rethinking Extractivism from Bolivia. *Latin American Perspectives* 46 (2): 141–160.

Mendoza Torrico, Fernando, and Félix Patzi González. 1997. *Atlas de los ayllus de Norte de Potosí, territorio de los antiguos charka.* La Paz: Comisión Europea Delegación en Bolivia.

Mumford, Jeremy Ravi. 2012. *Vertical Empire: The Struggle for Andean Space in the Sixteenth Century*. New Haven, CT: Yale University Press.

Murra, John V. (1976) 1985. "The Limits and Limitations of the 'Vertical Archipelago' in the Andes." In *Andean Ecology and Civilization,* edited by Shozo Masuda, Izumi Shimada, and Craig Morris, 15–20. Tokyo: University of Tokyo Press.

Oporto Ordóñez, Luis. 2007. *Uncía y Llallagua: Empresa minera capitalista y estrategias de apropiación real del espacio (1900-1935).* La Paz: IFEA/Plural Editores.

Perreault, Tom. 2013. "Dispossession by Accumulation? Mining, Water and the Nature of Enclosure on the Bolivian Altiplano." *Antipode* 45 (5): 1050–1069.

Platt, Tristan. 1982. *Estado boliviano y ayllu andino*. Lima: IEP.

Platt, Tristan. 2009. "From the Island's Point of View. Warfare and Transformation in an Andean Vertical Archipelago." *Journal de la Société des Américanistes* 95 (2): 33–70.

Postero, Nancy, and Nicole Fabricant. 2019. "Indigenous Sovereignty and the New Developmentalism in Plurinational Bolivia." *Anthropological Theory* 19 (1): 95–119.

Poveda Ávila, Pablo. 2014. *Formas de producción de las cooperativas mineras en Bolivia.* La Paz: CEDLA.

Querejazú Calvo, Roberto. (1977) 1998. *Llallagua: Historia de una montaña.* La Paz: Libros Maravillosos.

Rivera Cusicanqui, Silvia. 2003. "El mito de la pertenencia de Bolivia al 'mundo occidental'. Requiem para un nacionalismo." *Temas Sociales* 24: 64–100.

Rivera Cusicanqui, Silvia. 2015. "Violencia e interculturalidad: Paradojas de la etnicidad en la Bolivia de hoy." *Telar: Revista del Instituto Interdisciplinario de Estudios Latinoamericanos* 15: 49–70.

Rivera Cusicanqui, Silvia. 2018. *Un mundo ch'ixi es posible: Ensayos desde un presente en crisis*. Buenos Aires: Tinta Limón.

Rowe, John Howland. 1957. "The Incas Under Spanish Colonial Institutions." *The Hispanic American Historical Review* 37 (2): 155–199.

Sanjinés C., Javier. 2004. *Mestizaje Upside-Down: Aesthetic Politics in Modern Bolivia.* Pittsburgh: University of Pittsburgh Press.

Sheller, Mimi. 2018. "Theorising Mobility Justice." *Tempo Social, Revista de Sociologia da USP* 30 (2): 17–34.

Sheller, Mimi, and John Urry. 2006. "The New Mobilities Paradigm." *Environment and Planning A* 38 (2): 207–226.

Silverblatt, Irene. 1987. *Moon, Sun, and Witches: Gender Ideologies and Class in Inca and Colonial Peru.* Princeton, NJ: Princeton University Press.

Tandeter, Enrique. 1992. *Coacción y mercado: la minería de la plata en el Potosí colonial, 1692-1826.* Buenos Aires: Editorial Sudamericana.

Tockman, Jason, and John Cameron. 2014. "Indigenous Autonomy and the Contradictions of Plurinationalism in Bolivia." *Latin American Politics and Society* 56 (3): 46–69.

Webber, Jeffery. R. 2011. *From Rebellion to Reform in Bolivia: Class Struggle, Indigenous Liberation, and the Politics of Evo Morales*. Chicago, IL: Haymarket Books.

3 Mining, infrastructure, and mobility in the Andes

Gerardo Damonte, Julieta Godfrid and Ana Paula López

Introduction

Spatial mobility has always been part of large-scale mining. In the stories of the emergence and decline of large-scale mining operations that are recounted in books, articles, and documentaries, processes of immigration, emigration, and displacement often figure prominently (Gluckman 1961; Kruijt and Vellinga 1977; Nash 1979; Epstein 1981; Finn 1998). Typically, infrastructure in its various forms comprises part of these stories. Depictions of the construction or destruction of buildings, as well as the implementation of infrastructure networks such as roads or housing systems, repeatedly appear in photographs, videos, and local accounts of mining boom and bust periods. However, few studies have paid attention to the specific role that infrastructure plays in shaping spatial mobility.

This chapter focuses on how accommodation arrangements, viewed as a form of infrastructure, have shaped the spatial mobility of workers and displaced people over time in territories with large-scale mining projects in the Andes. We analyze the relationship between such infrastructure and spatial mobility patterns in two Andean territories: Espinar in Cusco, Peru, and Los Andes in Valparaiso, Chile, where the Tintaya-Antapaccay and Corporación Nacional del Cobre de Chile (CODELCO) copper mining projects are located, respectively. We examine the implementation of accommodation-related infrastructure and changing mobility patterns during two periods: (i) from the 1960s to the 1990s, when the predominant policies were focused on building mining camps and towns in which to house workers and displaced people in the context of an urbanization-based development paradigm; and (ii) from the 1990s to the present, a period in which infrastructure implementation policies have tended to be guided by a corporate social responsibility (CSR) paradigm, as a response to global demands for sustainable development.

In the context of extractive expansion, infrastructure has been studied in terms of its materiality and capacity to convey goods, resources, and people amid processes of change: as the material realization of progress, political dominance, or struggles (Ferguson 1999, 2012; Mitchell 2002; Joyce 2003; Baghel and Nüsser 2010; Anand, Gupta, and Appel 2018). Drawing on this literature, we explore the infrastructure that underpins the provision of labor and natural resources in two extractive operations in order to show how infrastructure shapes the temporality,

DOI: 10.4324/9781003313236-4

pace, and direction of movement of workers and local inhabitants. Following Larkin (2013, 328), we define infrastructure as built networks that facilitate the flow of goods, people, or ideas and that allow for their exchange across space. We focus primarily on the implementation of different accommodation arrangements – including houses, camps, and the provisioning of basic services – and explore the way in which they transform people's mobility.

Our argument is twofold. On the one hand, we contend that extractive development models, which are a function of technical–political projects to facilitate extractive operations, shape the design and implementation of infrastructure. On the other hand, we assert that social groups establish specific relationships with infrastructure as symbols, shaping the ways in which they appropriate it. Thus, while infrastructure serves as a substrate to people's spatial mobility, the infrastructure-mobility relation is also conditioned by the ways in which people appropriate, reinterpret, or contest corporate infrastructure designs in the context of social and political changes and conflicts.

Moreover, we frame infrastructure design within development models that serve technical–political projects oriented toward structuring people and territories on the basis of technical knowledge (Damonte, Godfrid, and López 2021). As a central part of these projects, infrastructures cannot be seen as neutral objects but as material changes that convey meaning to inform territorial, environmental, and social transformations (Collier 2011). As Harvey (2010, 2018) asserts, infrastructure generates both different expectations of development and the social space where actors struggle to consolidate their power. When projects are exclusionary, they prefigure scenarios of inequality, dispute, and conflict in which power relations between state, corporate, and local actors come into play.

Technical–political projects may encompass different (and disputed) development models that are expressed in distinct forms of infrastructure. In the twentieth century, companies generally built camps and towns to house mining workers, their families, and displaced people, and these urban spaces were presented as symbols of social progress and modernity (Epstein 1981; Finn 1998; Ferguson 1999; Miranda 2019; León Castro 2022). They were also fertile grounds for workers' political participation and struggles against corporate and state political dominance (Flores Galindo 1974; Nash 1979; Finn 1998; Damonte 2007). More recently, accommodation arrangements in large-scale mining have been mostly a function of new "outsourced" labor systems in which many workers are hired by contracting firms (Huallachain and Richard 1996). Companies in the twenty-first century still build accommodation infrastructure to compensate people displaced from mining expansion zones (Damonte and Glave 2019; León Castro 2022), but there has been a shift away from building towns to house workers and their families toward the construction of temporary accommodation facilities for workers while they are on site (Vega-Centeno 2011; Damonte, Godfrid, and López 2021; Damonte et al. 2022).

For the purposes of this chapter, we define mobility as the realm, temporality, motivation, and direction of people's routine movements in everyday life, in which we understand "realm" as the spaces that structure routine mobility and "temporality" as the ways people organize this mobility daily and seasonally (Jiménez 2009). To analyze this structured and routine spatial mobility in both our cases, we introduce the concepts of "short commuting" as the mobility of people within local spaces encompassing adjoining localities (Canales, Canales, and Hernández 2018), and "long-distance commuting" as forms of interregional mobility within a country (Jackson 1987; Houghton 1993; Aroca and Atienza 2011; Storey 2016).

This chapter draws on two case studies from the Andes to explore different ways in which housing infrastructure can shape and constrain the mobility behaviors of impacted communities and workers. We have chosen these cases because they involve territories with large-scale, open-pit mining operations where we have observed significant spatial mobility and urbanization processes and changes in relation to mining activities. Our analysis of the case studies draws on secondary sources such as public databases, reports, and archives, as well as primary data collected through ethnographic instruments such as in-depth and semi-structured interviews carried out in 2018, 2019, and 2022.[1] This chapter is divided into four parts. After this introduction, in the next two sections, we present the cases of Espinar in Peru and Los Andes in Chile. Finally, we present some final remarks as conclusions.

Peru case: Espinar, Tintaya-Antapaccay

Tintaya-Antapaccay, operated by the multinational mining company Glencore, is located in the district and province of Espinar, in the department of Cusco, Peru. The mine is situated more than 4,000 meters above sea level in the Andean mountain range. Traditionally, the province of Espinar was a predominantly rural area; in the 1981 census, it had a rural population of 36,510 and an urban population of just 8,029. But since 1982, with the development of mining activity, the situation has been reversed. Today, the population is predominantly urban, and most inhabitants are employed in the service sector. The mine has gone through different stages. In 1985, the state-owned Empresa Minera Especial Tintaya SA started operations on the right bank of the Salado River. In 1994, the state mandated the privatization of public companies and awarded the contract for the Tintaya mine to the US-based consortium Magma Copper Company. Since then, the mine has changed owners on four more occasions.

1981–1991: State, imposition, and concentrated mobility

The Tintaya mine started out under state control. After the nationalization of large-scale mining in 1971, the Peruvian government sought to link large-scale mining activity to a national sense of development. As stated in the 1971 Mining Law, "For the first time [the state] assumes effective control of the [country's] vast mining resources." One example of this disposition toward mining can be seen in the slogan "copper for Peru." Large-scale mining thus came to be attached to a nationalist technical-political development project. It was in this context that Tintaya was established, although it would later pass into private hands in 1994.

Tintaya promised to be not only a mining operation but also a development pole in the impoverished province of Espinar. The Tintaya mining camp, which was constructed to provide housing and basic services to the families of mining workers, most of whom came from other cities in Peru (particularly Lima, Arequipa, and Cusco), became a symbol of modernity insofar as it offered elements of an urban lifestyle that were then only available in Peru's major cities. Tintaya thus provided a developed urban space that set it apart from surrounding communities, which had limited or inadequate access to basic services such as sanitation, electricity, education, and healthcare. The camp infrastructure, with its symbolism of modern urban living, was intended to facilitate extractive development.

When the mining project required the expropriation of land, the state made plans to establish the mining town of Tintaya Marquiri, with all the services that the relocated population would require (León Castro 2022). For some residents, this decision was consistent with their expectations of progress, which they had come to associate with the characteristics already on display at the Tintaya mining camp. Regarding the expectations that plans to construct Tintaya Marquiri generated, one inhabitant of Tintaya Marquiri recalled:

> So, from there comes the agreement that says that we are going to be in a new city. . . . There they will have streets, squares, avenues, electricity, water, showers, beautiful streets. . . . Ah, we are going to be better off, we are going to be supported, we are going to build.
>
> (quoted in León Castro 2022)

In the end, however, the relocation was carried out by force, driven by business contingencies. As one displaced resident remembered:

> They threw us out, it's that simple, like a genocide. People didn't even want to leave. On many occasions we saw our parents: one moment they were having conversations, the next their houses were being demolished, with all their things inside. If you didn't leave, they threw you out anyway. They went in with their diggers, demolished, [while residents were] in tears.

For this interviewee, accommodation facilities for displaced people were a symbolic materialization of the violence exercised by the company (Appel 2012). Moreover, the new residential settlement of Tintaya Marquiri was far from what was promised. The infrastructure for housing families was deficient, with dirt roads, unfinished dwellings, and a lack of services.

In ensuing years, the mine's managers promised to turn the new settlement into a model community through home improvements and other development projects. Gradually, the infrastructure in Tintaya Marquiri improved. Between 1983 and 1990, the settlement was connected to electricity and drinking water networks, while a school and other services were provided. The residents, with the help of the company, took charge of managing the services that comprised the housing infrastructure. The memories of high-quality services then associated with mining activity linger in the minds of residents. As one inhabitant recalled about the schools of Tintaya Marquiri, "When the company was in charge of management [of the schools], good teachers were brought in, because the company brought in the highest ranked, the top levels, as they paid a certain amount and had an agreement with the Ministry of Education."

Housing for workers was also built in Tintaya Marquiri, creating an urban continuum with the mining camp that grew as the operation and the workforce expanded. In 1983, the mine generated an estimated 1,124 direct jobs, 80 of which went to community members (Lovón Zavala, Díaz Correa, and Echegaray Adrián 1984, 33). Although most of the relocated residents did not find stable work at the mine, many were employed on a temporary basis. For these relocated residents, who had come from rural backgrounds, this urban environment, with its service infrastructure dependent on mining activity, represented a new kind of living space. Residents who were unable to engage in mining employment had to migrate in search of work since herding and agriculture were no longer viable ways of life in the mine's area of influence.

The traditional destinations to which local inhabitants emigrated were Cusco and Arequipa: important cities where, in addition to work, it was possible to find service infrastructures considered superior to that available in Espinar, apart from the mining camp. However, over time, the neighboring town of Yauri assumed greater significance as a destination. On the one hand, it became a focal point for individuals looking for work at Tintaya, causing the population to increase exponentially in the first decades of mining: in 1981, the urban population at the district level was 6,113, and by 1993 it had grown to 18,545. On the other hand, Yauri's urban growth was also reflected in a notable improvement in its housing infrastructure. Yauri thus became not only a dormitory town for temporary immigrants in search of mining employment but also a pole that attracted nearby rural populations seeking to enter urban life to improve their living conditions. Urban living as a development model gradually gained ground, creating significant demographic and spatial changes in Espinar. Over the years, expectations of a better urban life – linked to this development model – were partially realized, although memories of state violence in the form of forced displacement lingered in local memory as a tragedy.

The new form of urban life resulted in the consolidation of certain mobility patterns. Mining prompted a new, constant flow of immigration to Yauri, as much from the cities of Cusco and Arequipa as from the surrounding rural areas. The pull factors of job opportunities and access to better housing infrastructure combined to render Yauri a thriving mining city. Mining workers organized their daily lives in the camp around the short daily trips they made to their workplace: the mine site. Among the resettled population that did not work permanently in the mine, Yauri and, at times, Cusco and Arequipa were part of their journeys. Mining employment and the opportunity to gain access to a life with "modern" urban infrastructure were key factors in defining mobility patterns for Espinar's inhabitants.

In the 1990s, the decade in which the Tintaya mine passed into private hands, the nationalist extractive discourse lost relevance. New corporations introduced labor-saving forms of production, while expanding mining operations made access to land and water in increasing quantities imperative for these companies. With the turn of the century, transnational corporate mining took on a dominant role in the country, accompanied by new local development discourses of CSR and a new development model based on the concept of sustainability.

1991–2022: Privatization, sustainability, and dispersed mobility

In the 1990s, Peru pivoted toward a development model that opened the economy to foreign capital, particularly extractive capital, which created the conditions for the privatization of state-operated mining. During this period, five mining companies acquired the Tintaya mine successively: Magma Copper Company in 1994; Broken Hill Proprietary (BHP) in 1996; BHP Billiton (product of the merger of BHP with Billiton PLC) in 2001; Xstrata Copper in 2006; and Glencore in 2013.

The new technical–political project proposed by the power elites of this period continued to have mining expansion at its heart, although without the discursive resources of extractive nationalism. However, mining expansion could not be achieved through the old formulas of work and urbanization. The more mechanized contemporary global mining industry has dramatically reduced its demand for labor, especially unskilled. This, together with the implementation of outsourced work systems intended to limit labor costs associated with employee benefits and social services, has significantly transformed companies' relationships with their environments: stable work has become increasingly scarce and the direct provisioning of accommodation infrastructure obsolete. Instead, the sector's new discourses of development have focused on the objectives of sustainability and CSR policies aimed at promoting forms of development that are not dependent on mining. Companies have promoted a discourse emphasizing their commitment to sustainable development and their strict standards of social responsibility in their interactions with communities.

For BHP Billiton, maintaining the old camp was no longer a priority. The workers, most of them now outsourced and temporary, were left to look for accommodation facilities outside the operation, primarily in Yauri. In Tintaya Marquiri, the company gradually stopped financing educational services, triggering a crisis in the local school, where enrollment plummeted from 700 to 200 students.

However, expectations of development based on the provision of jobs and urban infrastructure did not die away. Local communities demanded improved housing and services as part of negotiations over access to their lands for extractive expansion. The company's response was to introduce a new form of negotiation and compensation. Compensation replaced labor as the primary social relation between the company and surrounding communities (Filer 1997; Levancher and Le Meur 2022).

Conflicts between communities and BHP Billiton were common. This situation eventually gave rise, in 2002, to a framework agreement between the company, the local government, and the communities. Under this agreement, the company took possession of the properties it required for extractive expansion and negotiated a social license to operate. In exchange, it was required to acquire land for members of Tintaya Marquiri and other affected communities as compensation for the forced displacement of the past, in addition to allocating a percentage of its profits to a local development fund (De Echave et al. 2009). Housing infrastructure was a key plank in these negotiations.

The company acquired the Jayuni, Ccopachullo, and Challcha properties located in the province of Espinar, and the Buenavista property in the neighboring department of Puno. These lands were earmarked for the inhabitants of Tintaya Marquiri, prompting great expectations of development among the population, largely because the mining company promised to provide housing, access to telephone services, basic sanitation, agricultural projects, among other amenities. However, these promises have not materialized, engendering feelings of frustration, disappointment, and annoyance among inhabitants.

One critical point in the land relocation process was access to housing. Relocated people from Tintaya Marquiri complained about the lack of access to decent housing in the initial years after moving. This pressured the mining company to build houses over the last decade. As one interviewee asserted, "We've suffered completely. . . . [The company] has only created a ranch for cattle, 36 little apartments. That's all well and good, but I would have liked that from the beginning. My children have suffered; they've eaten in dirt."

As in the past, when the mine was state owned, the power of infrastructure as a symbol of progress was once again a determining factor in displacing families from the community. This time, however, private mining companies took precautions to avoid imposing infrastructure. Following CSR policies, the company opened negotiation spaces for the design of new housing infrastructure for the displaced. The plan for housing design was submitted to the community and land authorities for their approval, with each 50-square-meter house having two bedrooms, a living room, dining room, and a bathroom, as well as a pair of solar panels for electricity generation and a biodigester as a sanitation system. One key community request was for a stable alongside each dwelling. These characteristics reflect resident's aspirations and concepts of development: for them, the meaning of living there is bound up with the ability to rear their cattle, sheep, and alpacas.

A paradigmatic example of the symbolic importance of infrastructure can be seen in the Villa Alto Huarca housing complex. The communal lands located in an area that is now the north pit of the Antapaccay mine were ceded by the Alto Huarca peasant community to the mining company in exchange for construction of the 350-house Villa Alto Huarca complex, along with a yearly community allocation of five million soles (today the equivalent of roughly US$1.35 million), scholarships, job placements, three annual health campaigns, and sustainable development projects (Exchange-Alto Huarca Agreement). In 2012, the houses in Villa Alto Huarca, located in the southeastern part of Yauri's urban periphery, were handed over to residents. In addition, Villa Alto Huarca has a school that provides primary and secondary education, a church, and a sports field.

The Villa Alto Huarca project was very well received by community members, among whom there had been a combination of optimism and pessimism about the prospects of the mine accepting this demand. The idea came from a returning community leader, who was inspired by the experiences of other mining sites in Peru and abroad. However, there were also complaints related to the poor quality of the construction materials and the need to limit buildings to two floors, a constraint dictated by the nature of the foundations that had been laid. At first, very few beneficiaries lived in the houses: just seven families moved there in the first year.

Over the years, a significant percentage of owners rented out or sold their homes, despite an internal agreement not to sell to people from outside to the community. Today, almost one-third of residents come from outside the community, and several houses remain unoccupied.

This story of Villa Alto Huarca illustrates how the empowerment of local leaders translates into negotiation processes related to infrastructure design. Thus, community members appropriate the very process of infrastructure building. This appropriation is both utilitarian and symbolic. Although the company builds these homes as residences for displaced families, the families re-signify them as symbols of progress and as economic assets. The result is that the homes no longer necessarily define one's place of residence when they can be sold or rented out. This freedom of movement also applies to temporary workers who are no longer anchored to the mining camp. Most workers labor under rotation systems – for example, 14 days on and seven days off – and spend their off days in different places, such as Yauri, Arequipa, or Lima. Those who work for sub-contracted companies spend their nights in hotels in the city of Yauri. This dual dynamic of job insecurity and political–economic empowerment for displaced families, expressed in new ways of relating to housing infrastructure, has defined mobility patterns in this period (see Figure 3.1).

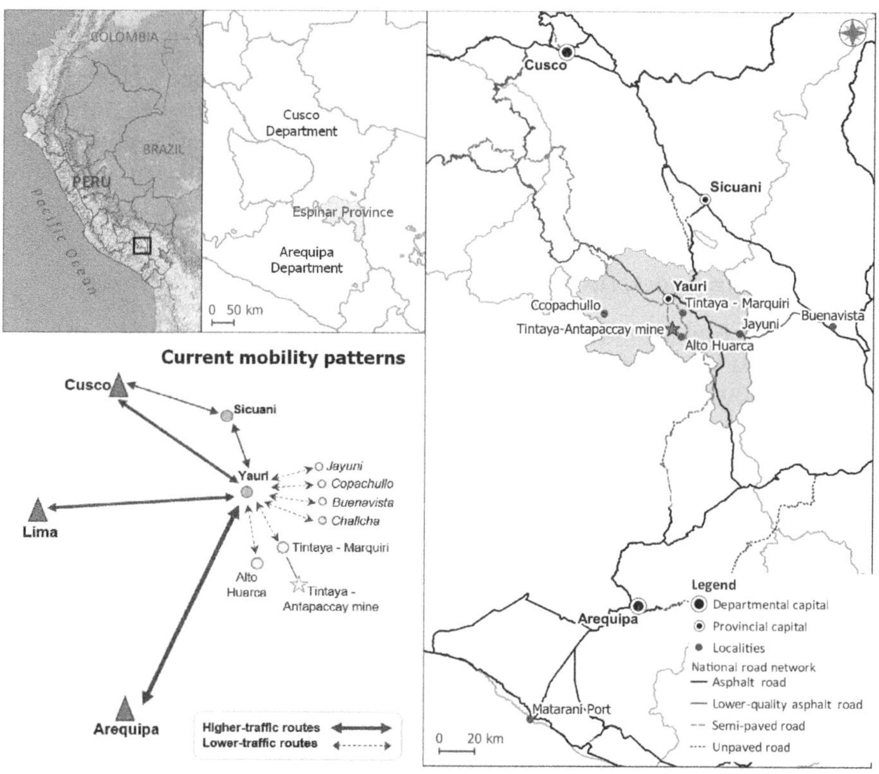

Figure 3.1 Current mobility patterns, Peru case. Figure by Ana Paula López Minchán.

The most notable changes in housing infrastructure during this period (1991–2022) were: the abandonment of the mining camp, the infrastructural crisis in Tintaya Marquiri, the construction of houses in rural communities, and the consolidation of Yauri as the main service-providing center in the region. These changes shaped the emergence of a new pattern of daily regional mobility in which three changes from the previous period can be observed.

First, the work commute is no longer from the camp to the operation but includes travel from Yauri and, to a lesser extent, Tintaya Marquiri. Second, there has been an increase in short commuting between rural areas and Yauri, with migration to this settlement motivated by expectations of mining work and the possibility of accessing better housing infrastructure. This has caused not only the increased urbanization of the district but also a growth in the number of families with dual residence: in Yauri and their communities of origin. As Castillo Guzmán (2020) shows, double residence facilitates flows of capital and people; in Espinar, this phenomenon has boosted mobility between Yauri and its rural hinterlands. Likewise, the logics of compensation introduced by company CSR practices sparked return migration to people's places of origin in order to take part in these negotiations with the aim of accessing more land. Third, long-distance commuting is also observable, in particular with some workers who, following a 14-days-on-seven-days-off work schedule, travel to distant cities such as Arequipa or Lima during their days off.

In summary, this case shows how accommodation infrastructure shapes mobility in different ways: mining housing as a promise of progress and a better way of life, with routine commuting between the mining camp and operations; infrastructure as an expression of violence framing the mobility of displaced people, such as the case of Tintaya Marquiri; and finally, the new corporate logics of negotiation and compensation shaping new flows of capital and people.

Chile case: Los Andes, División Andina

The División Andina project, operated by CODELCO, is located in the province of Los Andes, Valparaíso region, Chile. The mine is situated in the Andes at an altitude of 3,000–4,200 meters above sea level, 54 kilometers east of the commune of Los Andes, which is one of the four communes that make up the province of Los Andes. In terms of distance, it is also 80 kilometers northwest of Santiago de Chile and 140 kilometers from Valparaíso. Because of its geographical location, the mine is mainly connected with the province of Los Andes, which had an estimated population of 110,602 in 2017 (Instituto Nacional de Estadística 2017), and the commune of San Felipe (estimated population of 76,844) in the adjoining province of San Felipe. The provinces of Los Andes and San Felipe are located in the Aconcagua Valley, and the configuration of agricultural and urban settlements has been closely related to the course of the Aconcagua River (Cabezas Ferrer 2021).

The primary driver of mining growth in the region has been División Andina. Exploitation has taken place in different stages. Construction of the mine began in the late 1960s, and in the 1980s, operations commenced at the Río Blanco underground mine, from which some 70,000 tons of copper were extracted annually. Then, in 1983, the South–South pit was opened, allowing Andina to reach an annual production of 110,000 metric tons, including 500 tons of molybdenum (Barros Garcés 1988). In 1987, exploitation began at the Don Luis open pit mine. Between 1992 and 1999, CODELCO implemented a US$ 1.1-billion investment plan to upgrade the Andina mine and increase its copper production capacity (Baros Mansilla 2010). The investment allowed the company to start producing more than 200,000 metric tons of copper annually, commencing in 2000. In 2013, the company presented the Andina 244 project, which sought to double its daily extraction capacity and reach annual production of 600,000 tons per year. However, the company was unable to move forward with this project as a result of mass social opposition, mainly from local producers and environmental activists concerned about the project's likely impacts on glaciers. CODELCO was forced to reformulate its proposal, named Traspaso Andina, and between 2014 and 2022 invested US$ 1,455 million to renew its operations. In 2022, Traspaso Andina was inaugurated, extending the useful life of the mine until 2052 ("Codelco inaugural Sistema" 2022).

1967–1990: State and concentrated mobility

Mina Andina was constructed between 1967 and 1970 by the private American company Cerro Corporation. During the construction stage, around 4,000 people worked on the project, most of them migrants from other parts of Chile (Baros Mansilla 2010). Since then, mining has functioned as a pole that attracts migrants from other parts of the country to Los Andes, for example from the commune of Lota in the Biobío Region (Pérez Leighton, Sánchez Rubín, and Delso Páez 2021). In addition to the mine itself, the project infrastructure included the construction between 1966 and 1969 of a company town known as Villa Saladillo, located around five kilometers from the settlement of Rio Blanco.

In 1971, under the presidency of Salvador Allende, the Andina mine was nationalized. At that time, Andina was one of five mines included in the program to nationalize Chile's large-scale copper mining industry. Together, Andina, Chuquicamata, Salvador, El Teniente, and Exotica accounted for 78 percent of all copper production in Chile (Fortin 1975). In 1971, the Copper Corporation was created (Constitutional Amendment Law No. 17450) to manage the nationalized industries. Later, CODELCO was created by Decree Law No. 1350/1976 under the military dictatorship led by Augusto Pinochet (1973–1990).

The nationalization of copper was approved by the Chilean Congress by unanimous vote. During the debates in Congress, senators argued that copper would "bring the people health, culture, work, housing, and wellbeing" (the words of Ramón Silva Ulloa, in 1971). Allende declared the day of nationalization to be a day of "national dignity and solidarity." In the presidential speech of 1971, Allende presented nationalization as follows: "Fellow miners, hard workers of the red metal: once again I must remind you that copper is Chile's wage, just as the land is its bread. The Chilean bread will be assured by the peasants with their revolutionary conscience. The future of the country, the wage of Chile, is in your hands."

Mining nationalization was accompanied by a series of graphics that depicted the symbolic importance of this event. In the lithographs, produced in 1971 and 1972, copper is presented as the pillar underpinning Chilean society. One, for example, shows copper bars as the foundation on which the heterogeneity of Chileans is sustained. In this iconic lithograph[2] published by Vicente Larrea, Antonio Larrea, and Luis Albornoz, the image is accompanied by a paraphrased poem by Pablo Neruda that says: "Chilean copper, you are the homeland, pampas, and people, sand, clay, school, house of the resurrection, fist, marching order, parade, attack, wheat, struggle, grandeur, resistance." During those years, a narrative was forged around the idea of copper extraction as a provider of well-being (work, housing, health, and education).

For the operation of Mina Andina, the Villa Saladillo camp city was built close to the mine site, 39 kilometers from the city of Los Andes. In Chile, the so-called "copper cities" have been particularly flexible, adopting in each case a housing model suited to the topography and the environment (Garcés 2003). Saladillo had facilities to house 4,000 people, of whom 850 were mining workers (Baros Mansilla 2010). There were also workers who lived outside the camp and were transported to the mine site to start their shifts each day. According to interviewees, at that time, the mine ran three shifts: 4:30 am–4 pm; 1:00 pm–1:00 am; and 9:00 pm–8:00 am.

Villa Saladillo had houses for families and single workers, a hospital, a fire station, a shopping center with a supermarket, two schools, a church, and a recreation and sports center (with a cinema, newsstand, gym, sports courts, and swimming pool). The housing complex was stratified according to the position employees held in the mine: (i) workers were housed in one of four apartment blocks, named Cactus, Naranjos, Costanera, and Cipreses; (ii) middle-ranking supervisors were assigned to two-story houses, known as Garden; and (iii) managers and top-ranking executives enjoyed larger, detached houses, which were known as Staff (Peréz Leighton, Sánchez Rubín, and Delso Páez 2021). There was also a bus station for transportation to Los Andes (Baros Mansilla 2010). According to testimonies compiled by CODELCO, children of former workers described their life in the Saladillo camp as "idyllic" and "very pleasant because there was a lot of social life" (CODELCO 2022).

In addition to the Saladillo camp, the company began construction, in 1974, of the Villa Minera Andina in the commune of Los Andes, so that employees and their families could be relocated from the camp to the city of Los Andes. The Granja María Sylvia property was acquired and subdivided for the construction of the houses. The project unfolded over four stages between 1976 and the late 1980s: (i) 192 homes, three buildings containing 36 apartments each, and a kindergarten; (ii) 116 homes, four apartment buildings, and a complex that included a plaza and sports center; (iii) four buildings of 128 apartments each; and (iv) 64 houses (Baros Mansilla 2010). Workers had the option of buying the houses in installments and becoming homeowners.

Later, in 1984, División Andina also opened the La Gloria housing complex, and acquired homes in Curimón, located five kilometers from the city of San Felipe (Baros Mansilla 2010). Toward the end of the 1980s, the process of transferring workers to residential areas in Los Andes was complete, and Villa Saladillo was gradually abandoned.

During this first period (1967–1990), all CODELCO workers and their families received a series of benefits that largely guaranteed them access to healthcare, education, housing, and recreational infrastructure. Additionally, each year the company provided workers' children with all their school clothing, as well as Christmas gifts. Feelings toward the "mining family" were reinforced by annual parties in which workers from the company's different sectors participated in a parade through the town, each group wearing its representative clothing. Until the 1980s, workers' daily travel was limited to the walkable distance between the Villa Saladillo camp from the mine operations. But starting in the 1980s, daily travel expanded toward Los Andes after the housing infrastructure was moved into the city.

1990–2022: New mining, sustainability, and dispersed mobility

The second period was marked by a series of transformations in the development model and, in particular, in the form of mining production. These productive changes were associated with a range of factors, including technological changes and the democratic opening that coincided with a boom in large-scale private mining in Chile (Lagos et al. 2017). Emerging communication and mining technologies boosted extraction and processing volume, while also enabling changes in the circulation of goods and workers. These productive modifications were linked to changes introduced during the military dictatorship (1973–1990), such as labor legislation reforms that resulted in the flexibilization of hiring systems.

In 1994, CODELCO signed a "strategic alliance" with the state and labor unions, which was key to increasing mining productivity. In the early 1990s, in the context of democratic opening, two phenomena affected CODELCO's position as a copper producer: an increase in union conflicts and a loss of competitiveness vis-à-vis the multinational mining companies that were entering the market. CODELCO gradually changed its hiring systems, reducing the number of workers it employed directly, in favor of contracts with mining service providers. Thus, from the 1990s, the increase in volumes extracted was in inverse proportion to the number of workers directly hired by the company. Across CODELCO (taking into account all its divisions), production of copper in tons rose by 28 percent between 1994 and 1998, while the number of workers fell by 15.41 percent between 1993 and 1997 (Oropesa 2004).

These transformations marked the end of the "company town" and the beginning of a new mining development model in which the company would no longer provide its workers with either the housing infrastructure or the benefits they previously enjoyed. This new development model in the mining sector has three interrelated characteristics: (i) the subcontracting of personnel as "contractors,"

(ii) a high rate of daily worker mobility, and (iii) connections between the company and the local population occurring mainly by way of CSR programs that follow a logic of compensation.

On the first point, in the case of División Andina, an increase in contractors reflected the broader pattern of CODELCO's practices nationwide. In 1989, División Andina had 1,809 workers (Mesa and Kaempffer 2004) who produced 110,000 metric tons of copper annually (Baros Mansilla 2010). By contrast, according to information provided in CODELCO's annual sustainability reports, from 2000 onward, the division's annual production of metric tons of copper doubled, directly employed workers ranged in number between 1,400 and 1,500, and contractors totaled around 5,000. In the provinces of Los Andes and San Felipe, mining has posted its highest rates of employment growth since the turn of the century (Canales and Canales-Cerón 2016). Related to the mining labor demand, Los Andes populations show a high rate of masculinization and a floating population (Uribe Sierra, Mansilla-Quiñones, and Mora-Rojas 2022).

The benefits and income received by the company's directly employed workers are significantly greater than those received by contractors. Indeed, conflicts involving contractors at CODELCO have been recurrent in recent times. On several occasions – in 2012, 2013, 2018, and 2019 – they have disrupted operations through blockades of the mine's access road ("Trabajadores contratistas" 2012; "Trabajadores subcontratados" 2013; "Grupo de trabajadores" 2013; "Contratistas de Codelco" 2019). However, irrespective of the differences between the benefits received by one type of worker over another, mining wages are still significantly higher than those received by agricultural laborers in the Aconcagua Valley area (Canales and Canales-Cerón 2016).

The second point refers to the increase in the daily mobility of workers. Unlike the previous model, in which the population migrated primarily to the commune of Los Andes, this new stage has been characterized by an increase in short commuting as well as population growth in the commune's peripheral settlements (see Figure 3.2). By 2009, in the provinces of Los Andes and San Felipe, the commuting rate in the mining sector had reached 79 percent (Canales, Canales, and Hernández 2018). To illustrate this daily mobility, while in 2002 the ratio of cars moving between the communes of Los Andes and San Felipe was one-to-one, by 2012, for every car that went to San Felipe, there were another three heading in the opposite direction (Carroza Athens 2017). In 2002, 3,665 cars entered the commune of Los Andes per day; by 2012, the number had reached 6,612 (Carroza Athens 2017). These data reflect the increase in short commuting between the localities surrounding the mine and Los Andes, which, as mentioned before, is the closest settlement to the mine.

The use of nearby communes as dormitory towns for mining workers partly explains the dramatic increase in population and traffic in these communes. Between the 2002 and 2017 censuses, the population of Chile grew by 16.26 percent. However, in the commune of Los Andes, population grew by just 10.81 percent, while in the rest of the province, the figure was 20.62 percent (with growth concentrated in the communes of Rinconada, Calle Larga, and, to a lesser extent, San Esteban)

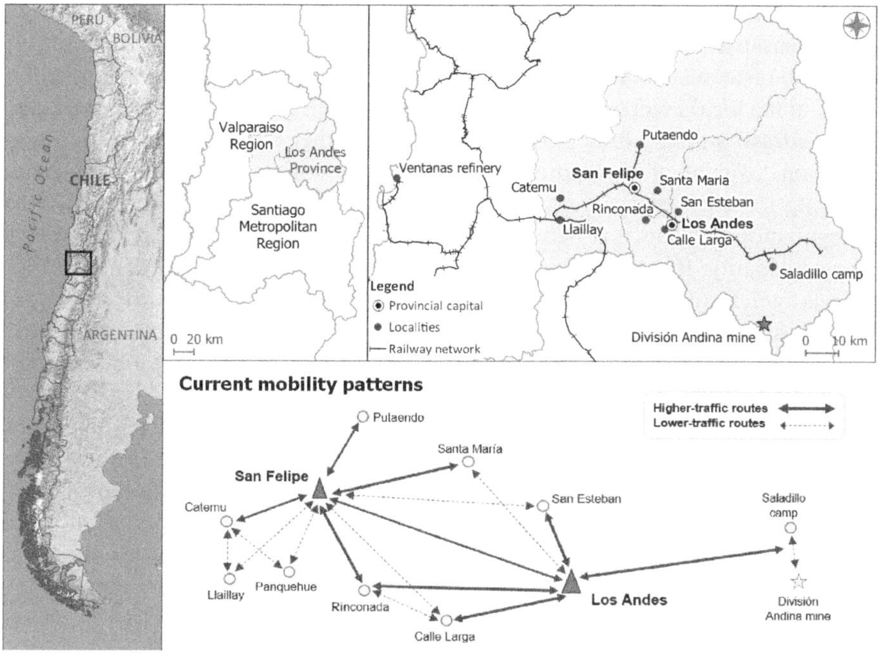

Figure 3.2 Current mobility patterns, Chile case. Figure by Ana Paula López Minchán.

(BCN 2023a). In the commune of San Felipe (San Felipe province), between 2002 and 2017, the population grew by 19.83 percent (BCN 2023b).

In certain parts of the city of Los Andes, as well as in other districts such as Calle Larga, the contractors' four-by-four trucks, usually red with yellow signage, are a constant presence. In some communes, such as Calle Larga, according to our interviewees, the large increase in vehicle circulation, and the extra pollution and noise that this has generated, has become a source of irritation to locals. In response, guided by a logic of compensation, CODELCO has launched social and urban infrastructure campaigns as part of its CSR programs.

Unlike the previous model, in which housing infrastructure had been built by the mining company, in the new period, housing growth has been largely the result of a set of state subsidies. In San Felipe, for example, between 2000 and 2020, around 10,000 people benefited from housing subsidies aimed at vulnerable sectors, as well as subsidies targeting the middle classes (Bravo and Cruz 2021). In San Felipe, the number of houses increased from 9,079 in 1982 to 23,908 in 2017, and the number of apartment buildings from 419 to 3,206 over the same period. In parallel, substandard housing was decreasing in number. Urban growth has also sparked an increase in facilities such as supermarkets and health centers.

As to the third element of the current mining development model, it is worth noting that the relationship between the company and the local population has also been reconfigured. The relationship between the company and the "mining family" is no longer direct but channeled through CSR programs under the logic of compensation. Through these programs, the company establishes a relationship with local residents whom they classify as "stakeholders." They offer "competitive funds" at the local level to finance infrastructures such as rural drinking water, irrigation infrastructure, public lighting, and various other community initiatives like educational, social, or sports programs. Though the company continues to provide financing for these measures, the type of infrastructure to be financed is subject to negotiation mechanisms between the company, residents, and the corresponding municipality. In the mining sector, this type of negotiation and compensation logic is typical of the implementation of CSR mechanisms (Ballard and Banks 2003; Haslam 2018; Maher et al. 2019). Two examples of infrastructure built or maintained under CSR programs help to explain the dynamics of mobility in the new mining model.

The first example is the improvement of infrastructure for the provision of rural drinking water. Within the framework of its CSR initiatives, in the last ten years, CODELCO has financed a program for improving water provision in those localities that are considered by the company to be within the mine's area of influence and where stakeholders live. This includes a variety of localities, as shown in Figure 3.2. The CSR water program has included well construction, pipeline upgrades, water pump purchases, and the training of technical operators in charge of distributing drinking water in rural areas ("Codelco Andina contribuye" 2014; "Programa APR" 2019; "Somos Codelco" 2021, 8; "Camino Internacional: Comunidades" 2022). The repair or construction of water infrastructure for rural areas has made it possible for the population to settle or remain in peripheral urban areas and to commute daily to the main city of Los Andes in search of work or other services.

The second example refers to the development of urban infrastructure in small towns near the mine. In the vicinity of the road leading to the mine, there is a locality popularly known as "Camino Internacional" where residents, who in several cases work as CODELCO contractors, have negotiated local infrastructure projects with CODELCO to improve the living conditions of more than 50 families living in the sector. Negotiations between neighborhood organizations have sometimes taken place through roundtables in the framework of CSR programs, and at other times through direct actions such as road blockades to prevent trucks from entering the mine. As a result of the negotiations, residents of the area managed to have school bus stops established, as well as lighting, urban signage, and road upgrades ("Codelco Andina, vecinos" 2017).

In summary, the second period (1990–2022) has been marked by a series of structural changes in the mining model that are reflected in the type of mobility and infrastructure. In the new mining model, accommodation needs are covered by state programs, private resources, or compensation policies under the umbrella of CSR programs. Unlike the previous model, the company no longer guarantees an accommodation infrastructure for workers or their families. On the contrary, in the new model, the accommodation infrastructure is the subject of negotiations and depends on the logic of compensation and CSR programs for its implementation.

The practices and infrastructures developed under the CSR umbrella have given rise to new urban mobility dynamics in the Andes, such as daily short commuting between mining towns and peripheral small urban settlements.

Conclusions

Accommodation infrastructure linked to mining development has undoubtedly conditioned daily spatial mobility in our case studies. It is worth noting some important differences between the two cases. For instance, while there has been continuity in corporate ownership at Andina, Tintaya has changed hands multiple times; while the displacement and relocation of local people have been central to Tintaya's history, this has not been the case with Andina; and Tintaya is comparatively more remote than Andina. Notwithstanding these differences, both cases show how the gradual shift from concentrated service networks in camps to the establishment of accommodation services in cities and mining suburbs is directly associated with a dispersion of people's daily flows, as well as an expansion in the regional extent of this mobility. Indeed, accommodation arrangements, as infrastructure, constitute the substratum that structures everyday spatial mobility. In highlighting the importance of infrastructures in shaping people's mobility in large-scale mining contexts, we can better understand the changes in mobility patterns from limited everyday mobility in mining enclaves (Kruijt and Vellinga 1977) to the more fluid mobility dynamics in the mining sites of today (Castillo Guzmán 2020).

In political and symbolic terms, infrastructure can be understood as the conceptual materialization of technical-political projects that, in our cases, sought to promote mining extraction by providing companies access to workers and local natural resources. In the first period analyzed, the promotion of large-scale mining was imbued with strong nationalist symbolic meaning in both Peru and Chile. In both countries, the governments of the day sought to promote mining by appealing to nationalistic sentiments and promises of development, equating extraction with national projects of mass appeal. In this regard, infrastructures functioned as the materialization of promises of development (Anand, Gupta, and Appel 2018), such as in the construction of the Saladillo camp. In the case of Espinar, following Appel (2012), we also saw the materialization of violence present in the destruction of old infrastructures and the imposition of new ones. Infrastructures as powerful political symbols shaped immigration, urbanization, and displacement as mobility processes related to mining development.

When the needs of mining changed, it was the sustainability paradigm that came to frame the construction of new infrastructure. This has been linked more to accessing resources and achieving local social license than to accommodating a workforce, which had become unnecessary given new demands and work structures. In this new context, the logics of compensation have informed the implementation of CSR discourses and policies. Mining infrastructures are no longer expressions of social progress, but of the materialization of compensation policies (Levancher and Le Meur 2022). Worker mobility has changed with the introduction of short- and long-distance commuting, while accommodation services for a better life have had to be found in nearby localities or distant cities, rather than adjacent to the mine itself.

However, there is a tension between corporate construction plans and local expectations of mining. Current CSR policies aim to promote infrastructures that are considered more sustainable, while local expectations hark back to company promises of housing, services, and work. This tension reveals negotiations over the materialization, through infrastructure, of a development that seems elusive. Yet, as we saw in the case of Tintaya Marquiri, local populations have the capacity to negotiate, appropriate, and re-signify infrastructures (Kertzer 1988).

Infrastructure as social relations is the materialization of tensions and power struggles between state or corporate elites who seek to impose functional designs on extractive development and the local population that resists and negotiates this imposition in line with their own expectations. It is at the heart of these social relationships that we can glimpse the ways in which infrastructure becomes a dynamic substrate that shapes local daily spatial mobility.

Notes

1 This study was funded in part by the Agencia Nacional de Investigación y Desarrollo Chile, Fondo Nacional de Desarrollo Científico y Tecnológico (FONDECYT Postdoctorado N° 3200013); the Consejo Nacional de Investigaciones Científicas y Técnicas (CONICET) of Argentina (postdoctoral scholarship 2018-2020 at Instituto de Investigaciones Políticas Universidad Nacional de San Martín); and Ford Foundation.

2 The lithograph can be viewed in the digital archive of the designer Larrea: https://www.archivolarrea.cl/collections/afiches-solos/products/afiche-cobre-chileno.

References

Anand, Nikhil, Akhil Gupta, and Hannah Appel, eds. 2018. *The Promise of Infrastructure.* Durham, NC: Duke University Press.

Appel, Hannah C. 2012. "Walls and White Elephants: Oil Extraction, Responsibility, and Infrastructural Violence in Equatorial Guinea." *Ethnography* 13 (4): 439–465.

Aroca, Patricio, and Miguel Atienza. 2011. "Economic Implications of Long-Distance Commuting in the Chilean Mining Industry." *Resources Policy* 36 (3): 196–203.

Baghel, Ravi, and Marcus Nüsser. 2010. "Discussing Large Dams in Asia After the World Commission on Dams: Is a Political Ecology Approach the Way Forward?" *Water Alternatives* 3 (2): 231–248.

Ballard, Chris, and Glenn Banks. 2003. "Resource Wars: The Anthropology of Mining." *Annual Review of Anthropology* 32: 287–313.

Baros Mansilla, Maria C. 2010. *Río Blanco y Andina: Visión de chilenos* [Rio Blanco and Andina. Vision of Chileans]. Codelco-Chile. https://www.codelco.com/prontus_codelco/site/docs/20111107/20111107191236/libro_andina.pdf.

Barros Garcés, Raúl. 1988. *Amarás al cobre como a ti mismo* [You Will Love Copper as Yourself]. Santiago: Editorial Andres Bello.

Biblioteca Nacional del Congreso. 2023a. Los Andes reporte comunal 2017. https://www.bcn.cl/siit/reportescomunales/comunas_v.html?anno=2017&idcom=5301

Biblioteca Nacional del Congreso. 2023b. San Felipe reporte comunal 2017. https://www.bcn.cl/siit/reportescomunales/comunas_v.html?anno=2017&idcom=5701#:~:text=Seg%C3%BAn%20el%20Censo%20de%202017,respecto%20del%20Censo%20de%202002

Bravo, Mario, and Rodolfo Cruz. 2021. "Un cultivo inmobiliario. Proceso de Urbanización en la Macrozona Central: Caso San Felipe 1982-2017." Degree Report, Facultad de Ciencias Sociales, Universidad de Valparaíso.

Cabezas Ferrer, Antonia J. 2021. "Entre el río y la ciudad: Reanudando la trama en los márgenes del Aconcagua entre San Felipe y Los Andes." Master's thesis, Facultad de Arquitectura, Diseño y Estudios Urbanos, Pontificia Universidad Católica de Chile.

"Camino Internacional: Comunidades del Camino Internacional se Capacitan en Operación de Sistemas APR." *Aconcagua News*, June 1, 2022. https://aconcaguanews.cl/portal/2022/06/01/camino-internacional-comunidades-del-camino-internacional-se-capacitan-en-operacion-de-sistemas-apr/

Canales, Alejandro I., and Manuel Canales-Cerón. 2016. "Población y territorio en el modelo económico chileno. El caso de la comarca del Aconcagua." *Papeles de Población* 22 (88): 9–46.

Canales, Alejandro, Manuel Canales, and María Hernández. 2018. "Trabajo y territorio en el nuevo agro chileno. Un estudio de *commuters* en tres comarcas del Valle Central." *EURE* 44 (131): 5–27.

Carroza Athens, Nelson. 2017. "Transformaciones en los patrones de movilidad cotidiana en nuevas geografías de la urbanización: El caso de la macrorregión central de Chile." *Cuadernos Geográficos* 56 (3): 207–227.

Castillo Guzmán, Gerardo. 2020. *Local Experiences of Mining in Peru: Social and Spatial Transformations in the Andes*. Abingdon: Routledge.

CODELCO. 2022. "La vida en Saladillo." Accessed June 19 2023. https://www.youtube.com/watch?v=mzBhaPPSGaY&ab_channel=ChileCodelco

"Codelco Andina contribuye al mejoramiento de la calidad de vida de los vecinos de Colina." *Codelco*, January 8, 2014. https://www.codelco.com/codelco-andina-contribuye-al-mejoramiento-de-la-calidad-de-vida-de-los

"Codelco Andina, vecinos de Riecillo y camino internacional impulsan mejoras en calidad de vida." *Codelco*, August 7, 2017. https://www.codelco.com/prensa/codelco-andina-vecinos-de-riecillo-y-camino-internacional-impulsan

"Codelco inaugura Sistema Traspaso Mina Planta, proyecto que dará continuidad a la División Andina." *Codelco*, January 19, 2022. https://www.codelco.com/codelco-inaugura-sistema-traspaso-mina-planta-proyecto-que-dara/prontus_codelco/2022-01-19/135855.html

Collier, Stephen J. 2011. *Post-Soviet Social: Neoliberalism, Social Modernity, Biopolitics*. Princeton, NJ: Princeton University Press

"Contratistas de Codelco Andina cortan Camino Internacional por mejores condiciones laborales en Los Andes." *El Observador*, November 5, 2019. https://www.observador.cl/contratistas-de-codelco-andina-cortan-camino-internacional-por-mejores-condiciones-laborales-en-los-andes/

Damonte, Gerardo. 2007. "Minería y política: La recreación de luchas campesinas en dos comunidades Andinas" [Mining and Politics: The Reenactment of Peasant Struggles in Two Andean Communities]. In *Minería, movimientos sociales y respuestas campesinas: Una ecología política de transformaciones territoriales*, edited by Anthony Bebbington, 117–162. Lima: IEP-CEPES.

Damonte, Gerardo, and Manuel Glave. 2019. "Reasentamiento involuntario: Políticas y prácticas en los Andes." *Mundo Agrario: Revista de Estudios Rurales* 20 (45): 1–15.

Damonte, Gerardo, Julieta Godfrid, and Ana P. López. 2021. "Mining and Urbanization: Ways of Generating Water Insecurity in Andean Territories." *The Extractive Industries and Society* 8 (3): 100954.

Damonte, Gerardo, Astrid Ulloa, Catalina Quiroga Manrique, and Ana P. López Minchán. 2022. "La apuesta por la infraestructura: Inversión pública y la reproducción de la escasez hídrica en contextos de gran minería en Perú y Colombia." *Estudios Atacameños* 68 (2): 1–32.

De Echave, José, Alejandro Diez, Ludwig Huber, Bruno Revesz, Xavier R. Lanata, and Martin Tanaka. 2009. *Minería y conflicto social* [Mining and Social Conflict]. Lima: CIPCA, CBC, IEP y CIES.

Epstein, Arnold. 1981. *Urbanization and Kinship: The Domestic Domain on the Copperbelt of Zambia 1950–1956*. London: Academic Press.

Ferguson, James. 1999. *Expectations of Modernity: Myths and Meanings of Urban Life on the Zambian Copperbelt*. Berkeley: University of California Press.

Ferguson, James. 2012. "Structures of Responsibility." *Ethnography* 13 (4): 558–562.

Filer, Colin. 1997. "Compensation, Rent and Power in Papua New Guinea." In *Compensation for Resource Development in Papua New Guinea*, edited by Susan Toft, 156–189. Monograph 6. Boroko: Law Reform Commission. Canberra: Australian National University, National Centre for Development Studies (Pacific Policy Paper 24).

Finn, Janet. 1998. *Tracing the Veins of Copper, Culture and Community from Butte to Chuquicamata*. Berkeley: University of California Press.

Flores Galindo, Alberto. 1974. *Los mineros de la Cerro de Pasco: 1900–1930* [The Miners of Cerro de Pasco]. Lima: PUCP.

Fortin, Carlos. 1975. "Compensating the Multinationals: Chile and the United States Copper Companies." *Institute of Development Studies Bulletin* 7 (1): 23–29.

Garcés, Eugenio. 2003. "Las ciudades del cobre: Del campamento de montaña al hotel minero como variaciones de la company town." *EURE Santiago* 29 (88): 131–148.

Gluckman, Max. 1961. "Anthropological Problems Arising from the African Industrial Revolution." In *Social Change in Modern Africa*, edited by Aidan Southall, 67–82. London: Oxford University Press.

"Grupo de trabajadores de la División Andina de Codelco cortan con barricadas el Camino Internacional." *La Tercera,* May 14, 2013. https://www.latercera.com/noticia/grupo-de-trabajadores-de-la-division-andina-de-codelco-cortan-con-barricadas-el-camino-internacional/

Harvey, Penelope. 2010. "Cementing Relations. The Materiality of Roads and Public Spaces in Provincial Peru." *Social Analysis* 54 (2): 28–46.

Harvey, Penelope. 2018. "Infrastructures In and Out of Time: The Promise of Roads in Contemporary Peru." In *The Promise of Infrastructure*, edited by Nikhil Anand, Akhil Gupta, and Hannah Appel, 80–101. Durham: Duke University Press.

Haslam, Paul A. 2018. "Beyond Voluntary: State–Firm Bargaining Over Corporate Social Responsibilities in Mining." *Review of International Political Economy* 25 (3): 418–440.

Houghton, Donald S. 1993. "Long-Distance Commuting: A New Approach to Mining in Australia." *Geographical Journal* 159 (3): 281–290.

Huallachain Breandan, and Richard Matthews. 1996. "Restructuring of Primary Industries: Technology, Labor, and Corporate Strategy and Control in the Arizona Cooper Industry." *Economic Geography* 72 (2): 196–215.

Instituto Nacional de Estadísticas. 2017. Censo 2017. https://www.ine.gob.cl/estadisticas/sociales/censos-de-poblacion-y-vivienda/censo-de-poblacion-y-vivienda

Jackson, Richard T. 1987. "Commuter Mining and the Kidston Gold Mine: Goodbye to Mining Towns?" *Geography* 72 (2): 162–165.

Jiménez, Marén. 2009. "Potencialidades de la medición de la movilidad cotidiana a través de los censos." *Notas de Población* (88): 163–184.

Joyce, Patrick. 2003. *The Rule of Freedom: Liberalism and the Modern City*. London: Verso.

Kertzer, David I. 1988. *Ritual, Politics, and Power*. New Haven, CT: Yale University Press.

Kruijt, Dirk, and Menno Vellinga. 1977. "The Political Economy of Mining Enclaves in Peru." *Boletín de Estudios Latinoamericanos y Del Caribe* 23: 97–126.

Lagos, Gustavo, David Peters, Alvaro Videla, and José Jara. 2017. "The Effect of Mine Aging on the Evolution of Environmental Footprint Indicators in the Chilean Copper Mining Industry 2001-2015." *Journal of Cleaner Production* 174: 389–400.

Larkin, Brian. 2013. "The Politics and Poetics of Infrastructure." *Annual Review of Anthropology* 42: 327–343.

León Castro, Camilo D. 2022. "Exploitation minière et relocalisation des populations au Pérou." PhD diss., École Des Hautes Études En Sciences Sociales.

Levacher, Claire, and Pierre-Yves Le Meur. 2022. "The Compensation Arenas in South New Caledonia. Minescape, Governmentality and Politics." *The Extractive Industries and Society* 11: 100999.

Lovón Zavala, Gerardo, Celso Díaz Correa, and Carlos Echegaray Adrián. 1984. *Tintaya, el desarrollo minero del Cusco* [Tintaya, The Mining Development of Cusco]. Cusco: Centro de Estudios Rurales Andinos "Bartolomé de las Casas."

Maher, Rajiv, Francisco Valenzuela, and Steffen Böhm. 2019. "The Enduring State: An Analysis of Governance-Making in Three Mining Conflicts." *Organization Studies* 40 (8): 1169–1191.

Mesa, Francisco, and Ana María Kaempffer. 2004. "30 años de estudio sobre ausentismo laboral en Chile: Una perspectiva por tipos de empresas." *Revista Médica de Chile* 132: 1100–1108.

Miranda, Fabio. 2019. "Gubernamentalidad corporativa y despolitización del reasentamiento de Morococha: Una aproximación etnográfica." *Debates en Sociología* 44: 91–117.

Mitchell, Timothy. 2002. *Rule of Experts: Egypt, Techno-Politics, Modernity*. Berkeley: University of California Press.

Nash, June. 1979. *We Eat the Mines and the Mines Eat Us: Dependency and Exploitation in Bolivian Tin Mines*. Nueva York: Columbia University Press.

Oropesa, Fabiola. 2004. "Diez años de alianza estratégica en Codelco Chile (1994–2004). De la confrontación a la cooperación y al diálogo." Degree Report, Universidad de Chile.

Pérez Leighton, Lorena, Triana Sánchez Rubín, and Miguel Delso Páez. 2021. *Los Andes: Puesta en valor de la identidad y patrimonio como eje de desarrollo sostenible*. Fundación Pro Cultura: Chile.

"Programa APR de División Andina beneficia sistema del sector Las Calderas de Calle Larga." *El Aconcagua*, May 16, 2019. https://www.elaconcagua.cl/2019/05/16/programa-apr-de-division-andina-beneficia-sistema-del-sector-las-calderas-de-calle-larga/

"Somos Codelco No 44." *Codelco*, 2021. https://issuu.com/micodelco/docs/rsc_44_a57c7acf8b7321

Storey, Keith. 2016. "The Evolution of Commute Work in the Resource Sectors in Canada and Australia." *The Extractive Industries and Society* 3 (3): 584–593.

"Trabajadores contratistas del cobre se toman ruta en reclamo por Mejoras salariales." *EMOL*, July 16, 2012. https://www.emol.com/noticias/nacional/2012/07/16/550867/trabajadores-contratistas-del-cobre-se-toman-ruta-internacional.html

"Trabajadores subcontratados de Codelco cortan tránsito en camino internacional Los Andes – Mendoza." *Diario U Chile*, May 14, 2013. https://radio.uchile.cl/2013/05/14/trabajadores-subcontratados-de-codelco-cortan-transito-en-camino-internacional-los-andes-mendoza

Uribe Sierra, Sergio, Pablo Mansilla-Quiñones, and Alejandro Israel Mora-Rojas. 2022. "Latent Rural Depopulation in Latin American Open-Pit Mining Scenarios." *Land* 11: 1342.

Vega-Centeno, Pablo. 2011. "Los efectos urbanos de la minería en el Perú: Del modelo de Cerro de Pasco y La Oroya al de Cajamarca." *Apuntes. Revista de Ciencias Sociales* 38 (68): 109–136.

4 Navigating gendered landscapes of mineral extraction

Spatial mobility, women's autonomy, and mining development in the Peruvian Andes

Gerardo Castillo Guzmán

Introduction

Despite promises of brighter futures, development projects and modernization processes have frequently led to uneven and contradictory forms of social and economic transformation (Ferguson 1990; Escobar 1995; Coronil 1997). Large-scale mining projects – with their tendency to globalize benefits and localize impacts – are major catalysts of social change among rural societies (Bebbington and Humphreys Bebbington 2018).[1]

As has been documented by many researchers, mining produces complex social, cultural, political, environmental, and economic changes which, in turn, can exacerbate existing inequalities (McMahon and Remy 2001; Rosser 2006; Bainton 2010; Damonte 2011; Loayza and Rigolini 2016; Arellano-Yanguas 2017). One aspect of inequality that has received particular attention from researchers is the differential impact of mining along gender lines, with a growing number of studies documenting how women have often been particularly disadvantaged by mining development (Lahiri-Dutt 2011; Soria 2012, 2017; Jenkins 2014; Horowitz 2017; Fent 2021).

This chapter is intended to contribute to this literature by bringing it into conversation with emerging research on the mining-mobility nexus. It uses a case study of a mining project in the northern Peruvian Andes (Rio Tinto's La Granja copper project in the region of Cajamarca) to explore how mining development fostered the spatial mobility and autonomy of women living in, or connected to, the project locality. The analysis concludes that, while women remained marginalized in absolute terms, they experienced some increase in spatial mobility, which in turn enhanced their capacity to act autonomously in physical, economic, and decision-making spheres. This chapter addresses what enabled these changes and what constrained them.

After this introduction, this chapter briefly reviews the literature on gender and extractive industries and provides a theoretical framework for understanding the relation between spatial mobility and power. The third section provides a brief account of the history and social characteristics of La Granja before mining development got underway around 1994. The fourth section explores how local women's mobility

DOI: 10.4324/9781003313236-5

practices have changed since Rio Tinto arrived. The chapter ends with a discussion of the possibilities and resistances that mining development generates regarding the spatial and social mobility of rural women in the Peruvian Andes and their capacity to enhance their physical autonomy, economic autonomy, and autonomy in decision-making (ECLAC-United Nations 2023). To do so, the chapter builds an explicit link between the theoretical lenses of the "new mobilities paradigm" (Sheller and Urry 2006) and the framework proposed by the United Nations (UN) Development Programme for understanding women's autonomy (United Nations 2010).

This chapter advances the argument that mining necessarily generates movement – movement of people, goods, ideas, aspirations, capital, technologies, pollutants, and so on. In generating movement, mining development also has the potential to weaken prevailing social and cultural barriers that limit women's physical autonomy; that is, their capacity to exercise their reproductive rights and halt gender violence. Greater physical autonomy can enhance economic and decision-making autonomy. Mining accelerates land transactions and the construction of housing and productive infrastructure, creates paid local jobs and small-business opportunities, and can challenge social structures – specifically kinship and conventional marriage and alliance norms – that hierarchically limit women's autonomy and capacities (e.g., owning land, getting paid jobs, moving freely). In brief, this chapter examines women's ability to physically and socially navigate the mineral extraction landscapes of the contemporary Andes. This navigation, although structured by gendered cultural norms, deviates from and challenges these norms and may potentially enhance women's autonomy, though not without resistance from men and female peers.

This chapter revisits material gathered during eight months of fieldwork in 2013, when local employment and procurement programs implemented by the mining company were at their highest levels. Ethnographic research included in-depth interviews with men and women from La Granja and surrounding villages, regional towns, and Chiclayo, the main destination city for migrants from La Granja, as well as situated observations of daily life routines. (For broader discussions of the research and its findings, see Castillo Guzmán 2015, 2020.)

In 2015, Rio Tinto decided to put the mining project on hold. In 2017, the company made a commitment to the Peruvian government to extend the project exploration period until the end of 2025. Despite the scaling back of project activity, some social programs remained active due to maintenance works and a social fund. In 2022, I updated the study, conducting further interviews with local villagers and company managers.

Navigating gendered mineral landscapes: a theoretical framework

Gendering extractive industries

Gender relations are a dimension of social life where contradictions between the global circulation of mining's benefits (e.g., profits) and the localization of its adverse impacts are dramatically experienced. Women are excluded from holding land rights in many mining localities and are particularly affected by deteriorating water quality and availability due to extraction. While the importance of gendered inequalities has long been acknowledged – for instance, in debates around migration and urbanization in the Zambian Copperbelt after World War II (Ferguson 1999; Damonte and Castillo 2011; Larmer 2021) – studies of the social and spatial transformations triggered by large-scale mining among rural populations traditionally have had a male bias, and many still do. However, a growing literature is exploring the multiple ways that mining interacts with gender relations and impacts the lives and experiences of both women and men.

These studies commonly show that the impacts – positive or negative – that mining produces are not gender neutral. For instance, as men take jobs in mining, they may withdraw from traditional subsistence activities, thus increasing the workloads of women, who take on more of these activities to maintain their families (Ward and Strongman 2011). As it is often harder for women to get jobs at the mine, they tend to be confined to agricultural work, where earnings are lower (Viale and Monge 2012). Weill (2021, 3–4) argues that conflicts between couples in mining contexts often concern the gender-related unequal distribution of the costs and benefits of mining activities – a kind of "masculinization of profits" and "feminization of losses." Imbalances in men-to-women ratios due to migratory flows in a male-dominated sector, along with deepening economic and social inequality between men and women and between miners and non-miners (Weill 2021), can also contribute to increased prostitution and alcohol consumption, with effects on women's lives, identities, and status (Jenkins 2014).

Even though mining activities have gender-differentiated impacts, social impact studies have not always included women in a differentiated manner, as a group of people with their own interests and expectations. Instead, these studies have tended to present local communities as homogenous (Soria 2017). Even more frequently, women have not been included in firms' social management plans, or in negotiation and consultation processes in territories adjacent to extractive projects (Cuadros 2010). Furthermore, when they do participate in negotiation and consultation, women may face political harassment and even overt gender violence from their partners and/or other men (Soria 2017). There is some evidence that agreements between mining companies and local communities are increasingly being used as a mechanism to facilitate meaningful engagement and sustainable development outcomes linked to gender equality and social inclusion (Keenan and Kemp 2014). However, corporate and/or local resistances still often exist. For instance, in her work among Kanak women in New Caledonia, Horowitz (2017) found that although women actively negotiate with and resist extractive projects, mining companies tended to reinforce women's subordinate social position and use that subalternity as a justification to dismiss their concerns. In short, the social and spatial transformations that mining development prompts in local communities do not occur in a social vacuum but are embedded in complex power dynamics that they may reinforce or help to challenge.

Politics of mobility

To understand the social fabric in which women's bodies move and, at the same time, to consider the interrelations between mobility practices and associated meanings, it is useful to focus on the interplay of three concepts: "mobility," "social navigation," and "autonomy."

Although mobility issues have been present in geographical research since the early twentieth century, scholarship traditionally focused on the physical movement of people and commodities from one place to another, through spaces regarded as closed, ahistorical, and depoliticized. The "new mobilities paradigm" (Sheller and Urry 2006) goes beyond physical movement to address mobility as a fragile "entanglement of movement, representations, and practices" (Cresswell 2010, 19) – or the getting from one place to another, the shared meanings of that movement, and the experienced and embodied practice of movement (Cresswell 2010).

The shift to the new mobilities paradigm places mobility in the social realm and so implies mobility subjects, mobility (and immobility) practices, and mobility politics – a politics which determines who has the right to move, why, and the ways that movement occurs. These politics are expressed in a mobility hierarchy – shaped by age, class, gender, ethnicity, and citizenship – which unveils structural differences regarding access to movement and the right to freely move (Cresswell 2010). In addition, as with space, mobility implies time: "It is the spatialization of time and temporalization of space" (Cresswell 2006, 4). Thus, to move freely – without fear and barriers – women need both space and time.

To help comprehend the mobility of La Granja's women amid social transformations in a changing environment, I borrow the concept of "social navigation" (Vigh 2009; Wijntuin and Koster 2018). "Navigation" – which literally means "to sail" – defines a special form of movement – movement within a fluid, changeable environment – that directs our attention "to the fact that we move in social environments of actors and actants, individuals and institutions, that engage and move us as we move along" (Vigh 2009, 420). Thus, "social navigation" refers "to the specific spatial and social practices of actors, how they actively move through, practice, cope with, seek to dominate, and learn how to live" (Wijntuin and Koster 2018, 4) in their environments, often in contexts of deep political, economic, and social inequality.

Finally, to illuminate the capacities of women to move freely in La Granja's social space, I borrow a framework developed by the UN. When the UN designed the millennium development goals for equality (2010), it proposed the operative concept of "autonomy" as an analytical and methodological tool to attend to the needs and demands of women. The levels of freedom that individual women can reach imply three interrelated dimensions of autonomy: (i) physical autonomy, which frees women from exclusive responsibility for reproductive and care tasks, allows them to fully exercise their reproductive rights, and precludes any form of gender violence; (ii) economic autonomy, which expresses women's capacity to produce and earn their own income and to control assets and natural resources; and (iii) autonomy in decision-making, which implies women's participation in their communities' decision-making in equal conditions with men (Soria and Castillo 2021). Public policies – and corporate policies, we may add – should allow and guarantee these three dimensions of autonomy, which are cornerstones for gender equality and parity citizenship (Benavente and Valdés 2014).

A gendered landscape in the rural Peruvian Andes

La Granja's history and society

The case study focuses on the La Granja project, one of the largest but not yet developed copper deposits in the world, located in Peru's northern Andes (see Figure 4.1). The project is owned by the multinational company Rio Tinto.

La Granja (which means "The Farm") is the village nearest to the mine and gives the project its name. Although situated in the region of Cajamarca, it is mainly connected to Chiclayo, one of Peru's main commercial cities, in the coastal region of Lambayeque, about 15 hours away by road.

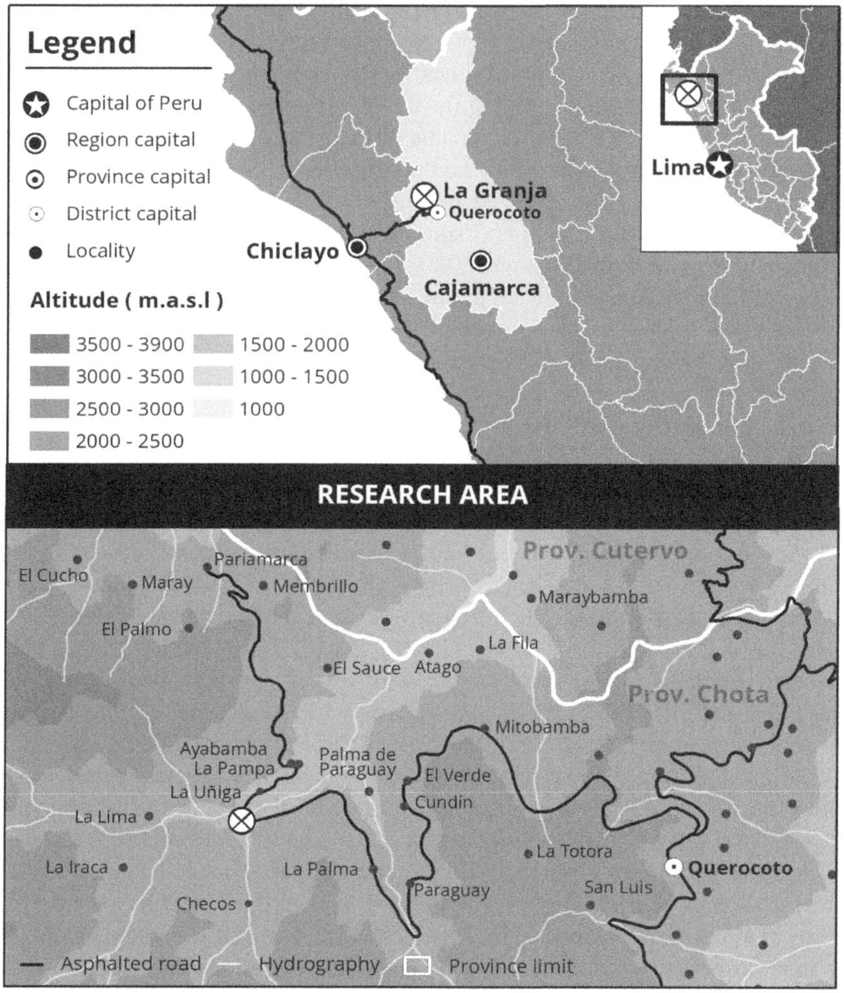

Figure 4.1 Map of La Granja, the study area. Elaborated Marilyn Ishikawa from Castillo (2015, 44).

Conversations held with older men and women indicate that La Granja's modern history goes back to the early twentieth century, when a large estate (*hacienda* in Spanish) was formed. The *hacienda* attracted the region's landless peasants, who settled with their families on the property and entered a pre-capitalist system of labor relations in which they were allowed to rent marginal lands from the landowner. Frequently, these peasant families wove personal bonds with the landowner, including ceremonial kinship ties through *compadrazgo* (godparenting), in the process building reciprocal, though unequal, relationships.

The landowner received payments – in the form of cash and in farming products – from the peasants in exchange for land, while also demanding services from men and women, including agricultural work (e.g., growing crops or grazing cattle) and domestic activities (e.g., in the manor house or weaving textiles). The landowner did not pay for these time-consuming services and expected them to be provided on-demand. He also at times forced tenants to hand over their young daughters for sexual favors, wielding the threat of expulsion from the *hacienda* as pressure to comply. What is more, tenants were not allowed to build their houses with brick and cement. Rather, they could only use mud and straw, so these constructions could easily be removed or burned. The landlord provided primary education to tenants' daughters and sons, while forbidding them from pursuing secondary studies (Castillo Guzmán 2020, 20–21).

Despite the servile and abusive nature of this system, the La Granja *hacienda* offered tenants a piece of land on which to work and settle and a degree of protection against crime, social violence, and destitution, in the context of an absent state (Castillo Guzmán 2020, 22). La Granja thus developed as a paternalistic and patriarchal model (Fuller 2001) of economic exploitation and social protection – embodied in the masculine figure of the *hacendado* (Deere 1990) – which pervaded social relations between genders. A local woman declared, without nostalgia:

> At *fiestas*, men used to drink and fight, and killed each other. I didn't like to leave my house. . . . I saw the fights from afar, and it scared me because they killed each other fighting. My husband was jealous when he was drunk.

This social dynamic was a perverse circle in which persistent poverty and land scarcity helped maintain the *hacienda* system, while that system reinforced poverty and sharply unequal social relationships. This situation did not go unchallenged: through passive resistance and active violent struggles, thousands of farmers defied the power of *hacendados* throughout the Andes for decades, until the 1969 land reform of Juan Velasco's military government brought the abusive pre-capitalist regime to an end. In La Granja, following land struggles between different sectors, the *hacienda* system ended, and its land was fragmented into dozens of small farms owned by independent rural families. The new owners began to build houses with more solid materials (mainly adobe and wood), usually near their farming plots.

A small village sprang up near the manor house. Poorly connected to the largest regional markets and with agriculture subsistence oriented, La Granja's population and economy barely grew. The small size of the plots made further subdivision difficult, and with limited job opportunities, many young people migrated to cities on the coast or in the Amazon.

As in many Andean societies (Bourque and Warren 1981; Deere and León de Leal 1981; Hamilton 1998), the land ownership system in La Granja offered only limited rights to women, who did not typically inherit or own their own land if they were single. Rather, land ownership usually only became possible for women once they married, thus tying their ability to access, control, and own land (the basis of subsistence) to marital status. (Of the adult women of the 14 families studied during my 2013 fieldwork, only two had owned property while single.) More generally, women were considered to be dependent on men – first their fathers, then their husbands – and only moved away from the parental home after marrying. The exceptions to this rule were single mothers and widows, who could obtain individual land rights. This gender-differentiated land-rights system in La Granja was linked to the general confinement of women to the domestic sphere, where they performed domestic tasks like cooking and weaving (Castillo Guzmán 2020, 109).

In sum, within La Granja's pre-mining development agrarian society, women lived under a patriarchal system in which they were largely excluded from land ownership and spent their days working in the fields and pastures (especially grazing cattle) and at home (assuming domestic work), though these (re)productive tasks were, in general, neither socially recognized nor paid (Castillo Guzmán 2020, 62).

Mineral development and the reconfiguration of La Granja's social space

Mining exploration began in the area in the early 1980s, as part of a joint effort between CENTROMIN, the state mining company, and Sondi, a German company, that identified the copper deposit. As part of these efforts, Sondi constructed a mining camp and an unpaved road between La Granja and the district capital, Querocoto. Mining works were abandoned until 1994. This was when the Peruvian state granted Cambior, a Canadian company, a five-year concession for the project, in one example of a broader push toward mining-sector privatization during this decade. While Cambior's re-initiation of exploration activities provided jobs and income opportunities to area residents, the company also began to purchase land, and it implemented an ill-planned resettlement process that involved legal and physical threats to pressure families to sell land and resulted in widespread displacement. During this period, Alberto Fujimori's authoritarian regime sought to attract foreign investment and demonstrate the pacification of the country following years of violence sparked by the Shining Path insurgency. In this context, the government closed schools and local health centers in La Granja to coerce the population to leave. State officials also accused several local leaders opposing the resettlement process of terrorism and initiated legal proceedings against them (Castillo Guzmán 2020, 24).

The result was a diaspora: 150 of the roughly 250 families that lived in the larger area sold their land and emigrated, mainly to Lambayeque. In the town of

La Granja itself, only seven families refused to leave, and Cambior demolished the remaining houses. For many of the people who left, displacement meant the end of a way of life, involving the interruption of productive activities, working conditions, social relationships, spatial mobility, sense of place, trust, and neighborhood. Without support from the company, the living conditions of many displaced families declined, and they struggled to survive in the new scenarios in which they found themselves (Castillo Guzmán 2020, 24).

In 2000, Cambior sold its mining and surface rights to Billiton, a company that later merged with BHP. However, following technical and financial studies, the Anglo-Australian company concluded the project to be not viable. In the wake of this decision, the company offered to return land (on advantageous terms) to displaced families, and many returned to La Granja.

Then, in 2006, Rio Tinto obtained the mining rights for the project, initiated exploration activities, and established local employment programs. This created new opportunities for area families, in the form of relatively well-paying jobs (with wages three times higher than those in farming) and opportunities to form small businesses. It also raised expectations that landowners would be able to negotiate compensation for eventual resettlement (i.e., through the sale of land to the company). Together, these factors promoted an influx of returnees to La Granja, most of whom were relatives of those who had left during earlier migrations (Castillo Guzmán 2020, 28). As a result, the number of houses in La Granja skyrocketed from seven (following Cambior's resettlement process in the 1990s) to more than 160 at the time of my fieldwork in 2013. After years of decline and deterioration, La Granja began to thrive, with local families building new houses, expanding old ones, and petitioning Rio Tinto for social and recreational infrastructure. The Sunday market at La Granja began to compete with that of the district capital as the area's main commercial center (Castillo Guzmán 2020, 28).

During this period, it was mainly the male leaders of La Granja, who generally belong to the same extended family and live in the town center, who negotiated with the mining company over access to jobs, business opportunities, and other benefits. However, families living outside the center that wished to negotiate directly with the company called this dynamic into question and started to create autonomous localities with their own authorities. In this way, the settlements of La Lima, Checos, and La Uñiga were established, thereby accelerating and deepening the physical and social fragmentation of La Granja – a process that had begun several decades prior (Castillo Guzmán 2020, 28).

Although Rio Tinto put the project on hold in 2015, La Granja has not reverted to the community it was prior to mining development. Of course, company activities have declined since the "boom" period sparked by Rio Tinto's arrival, when the number of houses in the village rose spectacularly. The number of local jobs offered by the company has fallen to 80–100 per year, from 2,000 per year at their peak, and the hiring period has shortened from six to three months.[2] However, mining-driven transformations have continued to be felt and have generated an endogenous process. For instance, according to a company senior officer interviewed in 2022, it is estimated that currently 80 percent of La Granja's families maintain double residency, mainly in the city of Chiclayo, where children and teenagers follow their studies; commercial networks and trips from La Granja, Chiclayo, and Chota remain strong; and, though many local businesses have closed after a period of unusual prosperity, some have become independent of the mining company and have expanded to regional and even national markets. The COVID-19 pandemic hit Peru's cities hardest, and hundreds of people returned to the area from Chiclayo, despite Cajamarca's Regional Governor trying to block the roads.[3]

The next section explores the changes in physical and social mobility that women from La Granja both experienced in the context of mining development and helped to shape.

Mining in motion: women's mobility and women's autonomy

During fieldwork in La Granja in 2013, women, by their own account and from my observations, were experiencing unprecedented levels of freedom regarding spatial mobility and autonomy, even though men continued to enjoy greater mobility. According to most women interviewed, men often traveled locally, regionally, and nationally and could be away for relatively long periods of time, while traveling was more difficult for women, and their trips were shorter, which in turn impacted the nature of their social interactions. The main barrier to women's mobility was the sexual division of labor prevalent in La Granja, in which women are responsible for most of the work of social reproduction. Indeed, women interviewees commonly affirmed that it was difficult for to leave town for many days because they had to take care of children or elderly family members and carry out domestic tasks (e.g., caring for small animals and the family vegetable garden). Further, men continued to have dominant presence in public spaces, and for many women, daily mobility was still restricted to spaces that could be considered physical appendices to their houses (gardens, pastures, grocery stores, restaurants). The range of men's spatial mobility, in contrast, was greater and included their homes, the mining camp, farming plots, roads and streets, and the sports field in the town center. Given the constraints on women's spatial mobility in daily life, social interactions were, for many, restricted to peers of the same age and gender (Castillo Guzmán 2020, 76).

At the same time, however, women developed strategies to socially navigate this male-dominated landscape and constructed their own times and spaces for socializing, such as volleyball games, church meetings, and the assemblies of the women's *ronda campesina*.[4] It was also apparent during my 2013 fieldwork that women's voices and their access the public arena in La Granja were expanding, though it must be recognized that women's spaces and mobilities remained more limited and less prestigious than masculine ones. As an example, women played volleyball in the village sports field on "off" hours, when men were working, and usually not on weekends. And volleyball is a less prestigious sport than soccer, which is played by men. For example, during the time of my fieldwork, a soccer tournament was organized during the religious festival of the Lord of Miracles, the main village celebration. It attracted around 12 teams, including some from other provinces within Cajamarca as well as from the region of Lambayeque, awarding around US$1,300 to the winner and $650 to the runner-up. In contrast, women played in a volleyball tournament with four teams from the village and neighboring towns that awarded prizes of around US$100 and US$65 for first and second place (Castillo Guzmán 2020, 76). Church activities were another meeting point for women, who gathered to pray and to decorate the chapel with flowers. Men did not consider these activities especially important, though a man presided (and still presides) over the Lord of the Miracles Association, an important local organization, which owns several lots in town in addition to carrying out its religious functions. Finally, the *ronda campesina* – composed exclusively of men – had more members and power than the women's *ronda*.[5] This is partly because the women's *ronda* was not an "organic" organization but rather was created on Rio Tinto's initiative for organizing the recruitment of women for its local employment program (Castillo Guzmán 2020, 77). In this context, many adult women tended to interact mostly with their children and other family members and reported this leaving them feeling isolated, without opportunities to develop friendship and solidarity bonds with their peers. By contrast, men had more interactions with other men, through their work and leisure activities (e.g., drinking, playing soccer, betting on cockfights, listening to music, gathering in the streets), allowing them to build bonds of gender solidarity (Castillo Guzmán 2020, 77). As Fuller (2001) observes, "the streets" are a space where men can accumulate social capital and prestige and form/reproduce the "male brotherhood" (Fuller 2001). Furthermore, the binary opposition between feminine home and masculine workspace/street tends to be reproduced when men appropriate the public domain (Da Matta 2002).

Despite these considerable and persistent limitations, women in La Granja were experiencing higher levels of spatial mobility than decades prior. As a local woman explained:

> Years ago, young women did not travel alone, but only when accompanied. . . . They went to work as domestic workers on the coast, and to the homes of close relatives. When traveling alone, women had to request authorization from their parents, even if the woman was an adult, because she was still living in the parents' home. If the woman lived with someone, then she had to request permission from her partner.

By the time of my 2013 fieldwork, however, women had more opportunities and fewer restrictions to leave the village and go to different locations in the region. Some had developed an important migratory registry that included travels to cities outside the region, such as Lima, and even outside the country, for example, Buenos Aires. In this respect, they were following in the footsteps of thousands of Peruvian immigrants seeking better labor opportunities elsewhere (Castillo Guzmán 2020, 78).

By 2013, women also faced fewer risks and less resistance when they moved around and used public areas and village streets than previous generations. As a local woman pointed out: "In La Granja women go out freely to walk, it is calmer than before. There are fewer fights, fewer drunk men." During my fieldwork in

La Granja, it was also not a surprise to find groups of young women strolling the streets or occupying the sports field in the town center to play volleyball. In addition, although women interviewees noted security risks in Chiclayo, they stressed how that city offered the chance to find a paying job, make friends, cultivate relationships among peers, and walk through the streets and shopping centers without restrictions imposed by their families in La Granja. Younger women especially tended to talk of the area's rural towns as backwaters, with nothing to do but work on the mining project (Castillo Guzmán 2020, 100). As one woman in her early 30s pointed out:

> At first, I didn't like Chiclayo. All the streets seemed the same to me, and I got lost. But then I made friends at work. I would leave the house and we would go for walks. I didn't miss La Granja at all. When I went there, I got bored. Now I like La Granja more because there is more activity. In the past there was nothing, not even work. Only old people yearn for farming. Only old people like how La Granja used to be.

In general, while in interviews men tended to present idealized constructions of rural towns, La Granja included, women denounced the violence of these spaces and stressed how life in the city – though insecure in many ways – provided anonymity where they could socially navigate with greater freedom and exercise greater physical autonomy, away from the control of their parents and male partners.

My 2013 fieldwork also revealed that La Granja's society had become more open to women buying land or properties directly from their owners, which offered greater possibilities for women to enjoy economic autonomy, along with decision-making and physical autonomy. This is despite a variety of persistent barriers that women faced. For one, because women had relatively fewer opportunities to earn income than men, their ability to save was limited. Also, though legal barriers to acquiring land and property did not exist, cultural and structural barriers still prevented them from accessing land in the same degree as men. Opportunities to inherit land were also restricted, partly because land fragmentation made it difficult to continue subdividing properties.[6] As a result, while agriculture was increasingly becoming a woman's preserve, in what has been called the "feminization of the countryside" (Remy 2014), men still mostly retained land ownership. This, in turn, placed men in a better position to negotiate with the mining company over land access (Castillo Guzmán 2020, 109), thus underscoring the continued importance of land ownership in the way people – or men, to be more precise – participated in the village's sociopolitical life. Like Himley (2016) has noted in the case of Barrick Gold Corporation's Pierina mine in the region of Ancash, land ownership and property systems are central to accessing mineral resources *and* the distribution of mining's costs and benefits.

Peru's economic boom of recent decades has made men in Andean regions increasingly willing to leave agricultural activities, searching for better wages in other sectors. While higher education levels and the previous migratory experiences of men partially account for their growing participation in non-agricultural activities, power relations better explain why men tend to capture economic benefits and opportunities for wage labor. As in other cases in the Andes (Farrell et al. 2004; Himley 2011), women in the La Granja area have received fewer economic benefits than men from the three mining companies that have operated in the area, in part because they are for the most part not the legal landowners (Castillo Guzmán 2020, 62).

Nonetheless, despite persistent gender inequalities and barriers, the "feminization of the countryside" is challenging some of the power structures that inform gender relations. Though women's work in farming tends to be seen as "complementary" to the household economy (rather than central to household reproduction), fieldwork in 2013 in La Granja suggested that women were progressively appropriating some agricultural profits and raising their voices to achieve collective representation in the locality. Moreover, due in part to the anti-discrimination policies and practices implemented by Rio Tinto, women were included in the company's temporary employment program and formed the female *ronda* (noted above) to organize their participation (Castillo Guzmán 2020, 63).[7] They were mainly employed in cleaning and cooking and constituted nearly a quarter of the mining project's local workers. A significant proportion were single mothers or single women. This was not necessarily because of the company's decision to focus its local employment program on vulnerable individuals, but rather because childcare and housework limited women's abilities to engage in wage labor. As one village woman stated, "The work that the mine has is for single women. When women with families work in the mine, they neglect their housework and their children. I had to resign to take care of my family." Observations such as these highlight the ongoing salience of a gendered division of household labor and the hidden barriers that women face in accessing paid jobs,[8] and it is worth noting that, in 2013, sources in La Granja indicated that more men than women worked in Rio Tinto's local employment program by a ratio of seven to three. This was (again) in large measure because women faced difficulties in maintaining paid employment while still being responsible for social reproduction tasks. Though it was also because Rio Tinto simply offered fewer jobs to women (Castillo Guzmán 2020, 77). In 2022, a Rio Tinto official acknowledged that the company failed to directly engage with the women's *ronda* when local jobs were available. The company's social team communicated with the leaders of the (men's) *ronda* and assumed they would share information with women's organizations; something that clearly did not happen.[9] Finally, in addition to accessing work at the mine (though not at levels equal to men), at the time of my 2013 fieldwork, women were also increasingly participating (as workers and proprietors) in local family businesses, such as restaurants, grocery stores, bars, and laundries, though much of this work could be considered an extension of domestic tasks, and it tended to restrict their mobility to the local arena (Castillo Guzmán 2020, 63).

The clear – if, in important ways, limited – progress that women were making in La Granja in the realms of education and physical, social, and decision-making autonomy was not welcomed by some and may have contributed to an increase in tensions within and between genders (Castillo Guzmán 2020, 63–64). As a young woman who was born in the town but grew up on the coast and later returned to La Granja to work in a relative's restaurant stated, "I don't like going out with boys from La Granja. They are very sexist, and they like to hit women." These tensions also manifest in accusations of infidelity, divorce, and separation from the family, leveled by both men and women, against women who work for wages. For example, an elderly woman declared with indignation:

> Women who have their own salary or business earn more money; this has caused them to be more liberal, because they are unfaithful to their partners and get divorced. Since so many people come from outside, the women think their husbands are small things. They devalue them.

The explicit link made in this statement between women's economic autonomy and their sexual freedom and free movement among men points to the challenges that women face in a changing social environment like that of La Granja. One female interviewee reported that when she was 15 years old, her family's difficult economic situation led her to leave her hometown. She went to Chiclayo, where she found work and made a life as an independent woman with her own social networks. Some years later, she became engaged to a man from La Granja and had a baby. The man worked odd jobs in the city, drank heavily, did not help with childcare, and tried to control her life. The woman reported wishing to return to paid work and separate from her partner (Castillo Guzmán 2020, 64).

Within this landscape in motion, it would be misleading to consider women as passive subjects of external mining impacts and structural hierarchies. Instead, using the concept of social navigation, this chapter highlights how women actively mobilized personal and social resources. As Wijntuin and Koster (2018) have shown regarding Dutch Moroccan girls in the Netherlands, single women in reproductive age have greater restrictions on their physical and social mobility than young girls and married and older women. Strolling in groups of peers or favoring the anonymity of the city are some of the strategies that young women employ to counteract these restrictions. Analyzing women leaders' political participation in a mining region, Soria (2017) describes how these leaders seek to attend meetings outside their localities accompanied by a woman friend to avoid allegations of sexual misconduct from their female peers and men.

At the same time, it is important to consider the ongoing imprint of La Granja's patriarchal social system. Indeed, some women interviewed did not question this system and even reported missing some of its aspects that seemed to be eroding. They thought that the changes mining development had brought to the area had created social disorder, which was often expressed in fears about women's sexual behavior: prostitution, licentious lifestyles, marital problems, and separation and divorce (Castillo Guzmán 2020, 100). For example, a woman from La Granja asserted a direct association between the mining project, access to money, and marital breakdown, saying, "Women leave their husbands. They are dirty women. I don't like this. When the company came, as men come from other places, women mess with married men. They are lovers for the money."

What is ultimately at stake in these dynamics is how women and men from La Granja – and elsewhere – understand how relations between genders should work – that is, the moral precepts that regulate the "proper" way for men and women – their bodies, practices, and representations – to behave. Consequently, some villagers thought that mining development – which represented the spearhead of modernization in the area – would not only destroy the material landscape but also was eliminating existing social relations and the moral order among residents, and between men and women. In this context, some interviewees (men and women) expressed a yearning to return to an idealized, pre-mineral development place (a type of conservative agrarian utopia), not only in the sense of a better relationship with nature but also in the sense of a patriarchal society of collective work, solidarity, benevolent authority, and family values (Castillo Guzmán 2020, 101).

Final remarks: mining development, mobility, and women's autonomy

This chapter argues that by its very nature in a globalized economy, large-scale mining necessarily generates movement – of people, goods, capital, and representations – at different scales. Examining the movements generated by the complex history of a copper-mining project shows how women from a rural village in the Peruvian Andes put into practice strategies to navigate a fluid – although still male-dominated – social landscape. As they navigated this landscape, women achieved greater autonomy. In the physical domain, they were able to travel more freely, often to towns and regional cities, and use public areas in the village to gather among themselves. In the economic sphere, women – single or married – had greater access to temporary employment programs, paid jobs, and opportunities to manage small local business. Finally, in the decision-making domain, there was some evidence that women from La Granja were organizing themselves for collective action, such as through participation in the female *ronda campesina*.

Using the new mobilities paradigm, this chapter has attempted to show that women in La Granja have not simply moved from one place to another through ahistorical and depoliticized territories. Rather, their mobilities imply specific subjects and bodies and vary depending on the diverse factors and contexts that women socially navigate. For instance, the mobilities – and associated meanings – of relatively young women with urban experience vary from those of older women with stronger attachments to rural environments; women's daily movements from potato fields to the family house in the village are different from the weekly travels of middle-aged women from La Granja to coastal cities, where their children attend school. These are the specific mobility politics that have been strongly shaped by gender and age in La Granja.

These mobility politics reveal structural differences regarding access to movement and the right to freely move. An unequal sexual division of labor – with women mainly in charge of unpaid and undervalued reproductive tasks at the household level – is perhaps the most persistent structure that has limited women's mobility in La Granja, both physically and narratively. In a sort of mobility moral landscape, women's attempts to gain more autonomy are often regarded –by men and women – as a betrayal of their home duties as mothers, wives, or daughters.

Mining development in La Granja has given rise to new economic and employment opportunities. Combined with the arrival of corporate mandates and policies framed within liberal social development agendas, this increased social and spatial mobility for many women. This greater mobility and freedom – which is associated with stronger connections to the city and urban life – has challenged previous gender relations and hegemonic practices and discourses of male control over life and women's autonomy. At the same time, however, this challenge has often been met with resistance and open gender violence.

This chapter has highlighted the theoretical and methodological value of centering on mobility for understanding mining's transformations and the interplay of power relations (the mobility hierarchy dictating who moves, how, and when, as noted in the theoretical section above) and individual agency. Studying this interplay involves the reinstallation of a politics of daily life into the literature on mineral development landscapes. From this perspective, the daily actions of the women who were the focus of this study represent a valuable example of women's ongoing struggles to build more equitable and diverse spaces in which to freely exercise and control the mobility of their bodies.

Notes

1 My thanks to Laura Soria, Matt Himley, David Brereton, Julieta Godfrid, and Iva Peša for valuable comments on early versions of the text, and to Rosemary Underhay, who edited the manuscript with professionalism.

2 Interview with a Rio Tinto senior advisor on external affairs, 15 August 2022.

3 Interviews with residents in 2022.

4 The *ronda campesina* is a self-defense organization originally created to fight rural banditry and cattle-theft in the region. *Rondas* are composed only of men, except in some places, like La Granja, where a parallel *ronda* of and for women was created.

5 While the *ronda campesina* of La Granja includes around 230 men, the female *ronda* does not reach even a third of that number.

6 In some of the examined cases, the couple registers new land and properties in the name of their children, but only the male ones.

7 Rio Tinto is one the few mining companies in the world which explicitly addresses the gender effects of its operations and has designed and implemented policies and protocols to manage those effects (Rio Tinto 2009).

8 In some cases, families hired a person to help with the housework, usually a teenager or an older woman from a nearby hamlet considered "more rural." This practice tends to perpetuate and exacerbate inequalities among women along lines of age and ethnicity. In other cases, women engaging in wage labor continued to assume heavy work burdens at home or passed some tasks to their daughters (Castillo Guzmán 2020, 63).

9 Interview with a Rio Tinto senior advisor on external affairs, 15 August 2022.

References

Arellano-Yanguas, Javier. 2017. "Inequalities in Mining and Oil Regions of Andean Countries." *Iberoamerican Journal of Development Studies* 6 (2): 98–122. doi: 10.26754/ojs_ried/ijds.255.

Bainton, Nicholas. 2010. *The Lihir Destiny: Cultural Responses to Mining in Melanesia*. Canberra: ANU Press. https://www.jstor.org/stable/j.ctt24hd35

Bebbington, Anthony, and Denise Humphreys Bebbington. 2018. "Mining, Movements and Sustainable Development: Concepts for a Framework." *Sustainable Development* 26 (5): 441–449. doi: 10.1002/sd.1888.

Benavente, María Cristina, and Alejandra Valdés. 2014. *Políticas para la igualdad de género: Un aporte a la autonomía de las mujeres*. Santiago de Chile: CEPAL-Cooperación Española.

Bourque, Susan, and Kay Barbara Warren. 1981. *Women of the Andes: Patriarchy and Social Change in Two Peruvian towns*. Ann Arbor: University of Michigan Press.

Castillo Guzmán, Gerardo. 2015. "Transforming Andean Space: Local Experiences of Mining Development in Peru." PhD diss. University of Queensland. doi: 10.14264/uql.2015.958.

Castillo Guzmán, Gerardo. 2020. *Local Experiences of Mining in Peru: Social and Spatial Transformations in the Andes*. Abingdon: Routledge.

Castillo Guzmán, Gerardo, and Laura Soria Torres. 2011. *Report: Gender Justice in Consultation Processes for Extractives Industries in Bolivia, Ecuador and Peru*. Lima: Oxfam and Societas Consultora de Análisis Social.

Coronil, Fernando. 1997. *The Magical State: Nature, Money, and Modernity in Venezuela*. Chicago, IL: University of Chicago Press.

Cresswell, Tim. 2006. *On the Move: Mobilities in the Modern Western World*. New York: Routledge.

Cresswell, Tim. 2010. "Towards a Politics of Mobility." *Environment and Planning D: Society and Space* 28 (1): 17–31. doi: 10.1068/d11407.

Cuadros, Julia. 2010. *Impactos de la minería en la vida de hombres y mujeres en el sur andino: Una mirada desde el género y la interculturalidad. Los casos Las Bambas y Tintaya*. Lima: CooperAcción.

Da Matta, Roberto. 2002. *Carnavales, malandros y héroes: Hacia una sociología del dilema brasileño*. México D. F.: Fondo de Cultura Económica.

Damonte, Gerardo. 2011. *Dinámicas rentistas: Transformaciones institucionales en contextos de proyectos de gran minería*. Lima: Grupo de Análisis para el Desarrollo. http://bibliotecavirtual.clacso.org.ar/Peru/grade/20121109040224/30_damonte.pdf

Damonte, Gerardo, and Gerardo Castillo. 2011. "Presentación: una mirada antropológica a las industrias extractivas en los Andes." *Anthropologica* 28 (28): 5–19. doi: 10.18800/anthropologica.2010-sup.007.

Deere, Carmen Diana. 1990. *Household and Class Relations: Peasants and Landlords in Northern Peru*. Berkeley: University of California Press.

Deere, Carmen Diana, and Magdalena León de Leal. 1981. "Peasant Production, Proletarianization, and the Sexual Division of Labor in the Andes." *Signs: Journal of Women in Culture and Society* 7 (2): 338–360.

Economic Commission for Latin American and the Caribbean (ECLAC)-United Nations. 2023. *Gender Equality Observatory for Latin America and the Caribbean*. https://oig.cepal.org/en/autonomies

Escobar, Arturo. 1995. *Encountering Development: The Making and Unmaking of the Third World*. Princeton, NJ: Princeton University Press.

Farrell, Leanne, Payal Sampat, Radhika Sarin, and Keith Slack. 2004. *Dirty Metals: Mining, Communities and the Environment*. Washington/Boston: Earthworks and Oxfam America. https://earthworks.org/wp-content/uploads/2021/09/NDG_DirtyMetalsReport_HR.pdf

Fent, Ashley. 2021. "Gender in Extractive Industry: Toward a Feminist Critical Resource Geography of Mining and Hydrocarbons." In *The Routledge Handbook of Critical Resource Geography*, edited by Matthew Himley, Elizabeth Havice, and Gabriela Valdivia, 103–113. New York: Routledge.

Ferguson, James. 1990. *The Anti-Politics Machine: "Development," Depoliticization, and Bureaucratic Power in Lesotho*. Cambridge: Cambridge University Press.

Ferguson, James. 1999. *Expectations of Modernity: Myths and Meanings of Urban Life on the Zambian Copperbelt*. Berkeley and Los Angeles: University of California Press.

Fuller, Norma. 2001. *Masculinidades: Cambios y permanencias. Varones de Cuzco, Iquitos y Lima*. Lima: Fondo Editorial de la Pontificia Universidad Católica del Perú.

Hamilton, Sarah. 1998. *The Two-Headed Household: Gender and Rural Development in the Ecuadorean Andes*. Pittsburgh: University of Pittsburgh Press.

Himley, Matthew. 2011. "El género y la edad frente a las reconfiguraciones en los medios de subsistencia originadas por la minería en el Perú." *Apuntes. Revista de Ciencias Sociales* 38 (68): 7–35. doi: 10.21678/apuntes.68.618.

Himley, Matthew. 2016. "Mining, Property, and the Reordering of Socionatural Relations in Peru." In *Mining in Latin America Critical Approaches to the New Extraction*, edited by Kalowatie Deonandan and Michael Dougherty, 208–226. London: Routledge. https://library.oapen.org/handle/20.500.12657/54229

Horowitz, Leah. 2017. "'It Shocks Me, the Place of Women': Intersectionality and Mining Companies' Retrogradation of Indigenous Women in New Caledonia." *Gender Place and Culture: A Journal of Feminist Geography* 24 (10): 1419–1440. doi: 10.1080/0966369X.2017.1387103.

Jenkins, Katy. 2014. "Women, Mining and Development: An Emerging Research Agenda." *The Extractive Industries and Society* 1 (2): 329–339. doi: 10.1016/j.exis.2014.08.004.

Keenan, Julia, and Deanna Kemp. 2014. *Mining and Local-Level Development: Examining the Gender Dimensions of Agreements Between Companies and Communities*. Brisbane, Australia: Centre for Social Responsibility in Mining, University of Queensland.

Lahiri-Dutt, Kuntala, ed. 2011. *Gendering the Field: Towards Sustainable Livelihoods for Mining Communities*. Canberra: The Australian National University. https://library.oapen.org/handle/20.500.12657/33742

Larmer, Milles. 2021, *Living for the City: Social Change and Knowledge Production in the Central African Copperbelt*. Cambridge: Cambridge University Press.

Loayza, Norman, and Jamele Rigolini. 2016. "The Local Impact of Mining on Poverty and Inequality: Evidence from the Commodity Boom in Peru." *World Development* 84: 219–234. doi: 10.1016/j.worlddev.2016.03.005.

McMahon, Gary, and Felix Remy, eds. 2001. *Large Mines and the Community: Socioeconomic and Environmental Effects in Latin America, Canada, and Spain*. Washington, DC: The World Bank.

Remy, María Isabel. 2014. "¿Feminización de la agricultura peruana?" *La Revista Agraria* 14 (158): 8–9.

Rio Tinto. 2009. *Why Gender Matters: A Resource Guide for Integrating Gender Considerations into Communities Work at Rio Tinto*. Melbourne: Rio Tinto.

Rosser, Andrew. 2006. "The Political Economy of the Resource Curse: A Literature Survey." Working Paper 268. Brighton: Institute of Development Studies.

Sheller, Mimi, and John Urry. 2006. "The New Mobilities Paradigm." *Environment and Planning A: Economy and Space* 38 (2): 207–226.

Soria, Laura. 2012. *Estudio sobre comunidades campesinas de Espinar y proceso de diálogo, negociación y consenso en actividades de exploración minera*. Final Report. Lima: Oxfam and Societas Consultora de Análisis Social.

Soria, Laura. 2017. *Liderazgos femeninos en contextos mineros: caso Tintaya en Espinar*, Cusco. Tesis de maestría, Universidad Nacional Mayor de San Marcos.

Soria, Laura, and Gerardo Castillo. 2021. "Enfoque de género en la gestión ambiental del sector minero a gran escala desde la década de 1990 en el Perú: dislocaciones y fragmentaciones." In *¿Una oportunidad perdida? Boom extractivo y cambios institucionales en el Perú*, edited by Gerardo Damonte, Barbara Göbel, Maritza Paredes, Bettina Schorr, and Gerardo Castillo, 129–156. Lima: Fondo Editorial de la Pontificia Universidad Católica del Perú.

United Nations. 2010. *Achieving the Millennium Development Goals with Equality in Latin America and the Caribbean: Progress and Challenges*. Santiago de Chile: United Nations Publications.

Viale, Claudia, and Carlos Monge. 2012. "La enfermedad chola." *Quehacer* 185: 80–85. http://www.desco.org.pe/recursos/sites/indice/824/2438.pdf

Vigh, Henrik. 2009. "Motion Squared: A Second Look at the Concept of Social Navigation." *Anthropological Theory* 9 (4): 419–438. doi: 10.1177/1463499609356044.

Ward, Bernie, and John Strongman. 2011. *Gender-Sensitive Approaches for the Extractive Industry in Peru: Improving the Impact on Women in Poverty and their Families*. Washington, DC: World Bank. http://hdl.handle.net/10986/2516

Weill, Caroline. 2021. "'El dinero, machista lo vuelve al varón': Conflictos, separaciones y reconfiguraciones de las relaciones de género bajo la influencia de los proyectos mineros en Espinar (Perú)." *Revue Interdisciplinaire de Travaux sur les Amériques* [online] 14. http://www.revue-rita.com/articlesvaria14/el-dinero-machista-lo-vuelve-al-varon-conflictos-separaciones-y-reconfiguraciones-de-las-relaciones-de-genero-bajo-la-influencia-de-los-proyectos-mineros-en-espinar-peru-caroline-weill.html

Wijntuin, Patricia, and Martijn Koster. 2018. "Dutch-Moroccan Girls Navigating Public Space: Wandering as an Everyday Spatial Practice." *Space and Culture* 22 (3): 1–14. doi: 10.1177/1206331218794603.

Section II

Central and West Africa

5 Chasing gold

Technology, people, and matter on the move in Eastern Democratic Republic of the Congo

Philippe Dunia Kabunga, Simon Marijsse and Sara Geenen

Introduction

Gold is not fixed in the earth. When talking about their work, Congolese gold miners say that they are "chasing" or "hunting" the gold veins, implying that these are moving prey. But chasing can also signify "to drive out" spiritually. What is this matter the miners are chasing? What is the matter with gold?

Thinking about gold and mobility not only directs attention to the movement of people who are chasing it, implying more-or-less extensive forms of labor migration. It also questions the movement and agentic qualities of the matter itself. Miners, for instance, continuously reshape their subterranean workspace through various techniques and technologies, which are in themselves often a result of an intricate web of movement and transport fraught with contingency (D'Angelo 2022). Rivers can rapidly push alluvial gold downstream during the rainy season. On the other hand, miners contend that skilled diviners can deceive miners, stimulating movement through hearsay and mystical stories about "the multiplication" of gold veins in some faraway place. Here, we turn to the world of the occult. When production diminishes, gold mysteriously "disappears."

Drawing on years of ethnographic fieldwork combined with historical research, this chapter analyzes different kinds of motion in and around gold mines in Eastern Democratic Republic of the Congo (DRC). The perception of gold as a mere economic resource is questioned through a kaleidoscopic perspective, shifting between different movements that presuppose it: people, technology, the otherworldly. This allows us to move beyond an understanding of mobility restricted to people. Being *à la chasse* (on the hunt) for the gold veins entails the use of manual, spirited (such as charms, spells, and divinations), and mechanized attempts and techniques for making gold veins visible and accessible, and for making tailings valuable and profitable. These techniques are not static but malleable, as is the subsoil they seek to crack open.

In the academic literature, the geological fixity of gold has been questioned by critical resource geographers (Valdivia, Himley, and Havice 2021), invigorating relational analyses of resource making (Bridge 2009). A concept such as "resource assemblage" gives us some analytical leverage to navigate the multiple registers at

DOI: 10.4324/9781003313236-7

work (Li 2014). It requires us to describe the agency of gold-as-resource as one that is multiple: how it moves materially (as a result of extraction or through the movement of water), and how it is valorized differently through different historical and cultural registers (colonial, customary, occult, international market) (Mol 2002).

Thinking about assemblages as the coming together of elements – humans, nature, minerals, technology, culture – we consider mining to be always entwined with motion. What is *on the move*, however, is not restricted to people. And what motivates this mobility can in no way be reduced to people either. Residues and deposits that were once unsuitable, inaccessible, or unprofitable can become a valid source of extraction because of (new) people and (new) techniques, and the continuously changing sites of extraction themselves. These movements need to be considered in their socio-technical entanglements and against a wider historical backdrop of moving people, technologies, and materials. And when diviners explain they make gold disappear, it impels us to move toward questions of an ontological nature.

Mining and mobility

A conventional understanding of migration sees it as a unidirectional movement; that is, from one place to another. Recent critical studies in migration, however, have challenged this understanding (Schapendonk et al. 2018). They argue that migration should be understood as "non-linear" movement (Mainwaring and Bridgen 2016) and highlight circular aspects. In some cases, migration can be considered a "mobile mode of existence" (Dahinden 2010; Tarrius 2002; Van der Velde and Van Naerssen 2011). In this chapter, we follow this line of thinking and draw attention to the *connections* – in time and in space – that are made when people, technologies, and matter are on the move.

In research on mining in Africa, three bodies of literature have spoken to this same question. First, anthropological studies from the twentieth century focused a lot on labor migration in the colonial era. In southern Africa, for instance, government labor recruitment boards required chiefs to supply cheap labor from the rural areas (Harries 1994; Jeeves 1985; Moodie 1994; Read 1942; Richards 1932; Schapera 1947). Tens of thousands of migrant workers were compelled to do low-wage work in the underground mines while they settled in European-modeled urban centers. Nevertheless, links with their rural homelands remained strong. From the 1930s, mining companies, together with the colonial administration and the church, used segregation, paternalistic practices, and social engineering to "stabilize" and control the workforce (Van Onselen 2001). This was very prominent in the DRC's Copperbelt region (Cuvelier 2011; Dibwe Dia Mwembu 2001; Fetter 1973; Rubbers 2019). However, Peša and Henriet (2021) also draw attention to the histories of farmers, small traders, and tailors who did challenge, in multiple ways, the unified "model of modernity" imposed by mining companies.

Second, research on artisanal and small-scale gold mining (ASGM) has long acknowledged the mobility of miners, emphasizing that rural people are pushed toward the mines – either longstanding extraction sites or so-called "rush" sites – in search of livelihood opportunities (Banchirigah and Hilson 2010; Hilson et al. 2017;

Jønsson and Bryceson 2009; Maclin et al. 2017; Mkodzongi and Spiegel 2020). Some mine workers move seasonally, some subsequently move from site to site, some go back to a site where they previously worked. Recent forms of mechanization in ASGM have ushered in an emerging body of literature focusing on technological change (Hilson and Maconachie 2020; Lanzano 2020; Lanzano and Di Balme 2021; Massaro and De Theije 2018; Verbrugge et al. 2021). In previous work, we have shown how such technological changes not only imply changes in work rhythm, governance, and space but can also re-energize colonial memories and the demand for local modifications of externally induced technologies (Dunia Kabunga and Geenen 2022; Marijsse and Munga 2022). In sum, in underground or underwater workplaces, bodies *move* technologies and techniques, and technologies simultaneously instigate bodies to move and hunt for often already "touched" soils (D'Angelo 2019; Nystrom 2014; Pfaffenberg 1998; Rolston 2013). Here, mobility is not an end in itself, but a means, a necessary condition for accessing gold (Bolay 2016).

A third body of literature draws attention to occult or otherworldly practices surrounding mining sites. In studies on ASGM, tales surrounding toxic gold speak to an earlier body of literature that served to dispel an all-too-rigid dichotomy between modernity and witchcraft (Geschiere 1997). Here, it is not only stories about quick wealth that may scare or attract miners, as described in the cases of Madagascar (Walsh 2003), Burkina Faso (Werthmann 2003), Benin (Grätz 2003), and Sierra Leone (D'Angelo 2014, 2015), but there are also underlying ontological questions about the agentic qualities of gold when diviners proclaim that they have made gold disappear (Kohn 2013). In this chapter, we do not focus on the movement of gold through the supply chain; as it moves from extraction into processing, and then to places that remain invisible for the mine workers. We rather focus on the visible or hidden presence of gold at the place of extraction, which is surrounded by rumors, stories, charms, spells, and divinations. This hints at a different kind of mobility, namely the agentic qualities of the subsoil itself (and its minerals).

Methodology

As authors of this chapter, we have often been on the move ourselves, chasing data and insights on ASGM in different provinces in Eastern DRC. In writing this chapter, we reconsidered these insights from the vantage point of mobility. This chapter makes use of archival data, as well as ethnographic data collected in two research projects funded by the Research Foundation Flanders (FWO) between 2018 and 2022 (one on technological innovation in ASGM and one on informalization in global gold production); as well as several other projects in which the authors were involved between 2008 and 2022.

Locating the sites

Despite the multiple and fragmented origins – in a geographical and temporal senses – of the data on which we drew to write this chapter, we locate four mining areas that have profoundly shaped the analysis we present here (see Figure 5.1).

Figure 5.1 DRC, with locations of the four territories covered in the chapter. Map by Bossissi Nkuba.

Kamituga is a city with a population of more than 150,000, situated 180 km from Sud-Kivu's provincial capital Bukavu, in the territory of Mwenga. It hosts some of the largest gold mining sites in the province. The city was built by *Minière des Grands Lacs* (MGL), a colonial company that started extracting gold in the 1920s (Geenen 2015). Today, a large share of the population depends on ASGM, either directly working in the underground pits and alluvial sites, or indirectly in trade and services. According to local statistics, at least 40 percent of the mine workers are not born in the territory (Mairie de Kamituga 2022), where indigenous groups are Balega. Beginning in colonial times, MGL recruited workers from the wider region as well as the neighboring countries of Burundi and Rwanda.

Further inland, the territory of Shabunda is located about 345 km west of Bukavu. Various minerals, including gold, have been mined here since colonial times. The indigenous groups are also Balega. Today, many smaller and bigger mining sites are scattered across the vast territory, which is difficult to access, and where security is volatile. On the Ulindi, Lubimbe, and Lugulu rivers, dredges are in operation. Workers come not only from other Congolese provinces such as Kasaï-Central, Kasaï-Oriental, and Haut-Katanga but also from Angola, South Africa, and China. In 2021, the population of the town of Shabunda was estimated to be around 95,000 (Territoire de Shabunda 2019).

To the south, the town of Misisi is located about 350 km southwest of Bukavu, in the territory of Fizi. The town of Misisi is estimated to have a population of around 90,000 (Poste du secteur de Ngandja 2021). The indigenous populations in this region are Babembe, but there are other local groups – such as Bavira, Balega, and Bangubangu – as well as foreign investors in ASGM, such as Burundians, Tanzanians, and Chinese. The security situation around Misisi is also highly volatile, with the presence of several non-state armed groups.

The towns of Durba and Moku, finally, are located in the northeast of the country, in the territory of Watsa, in the province of Haut-Uele. Durba hosts the Congolese headquarters of multinational Kibali Gold Mines, but many people are working in ASGM too, including from neighboring provinces (Tshopo, Ituri, North Kivu, Bas-Uélé), and foreigners from the Central African Republic, Uganda, and South Sudan.

Connecting the dots

In these sites, we have conducted hundreds of interviews with many different categories of workers, whom we purposively sampled. These include pit managers, gold traders, drillers, transporters, machine and dredge operators, manual grinders, female stone washers, divers, cooperative leaders, civil society members, customary chiefs, government authorities, and many more. We have spent many days observing work and life in the mines, returned on several occasions, engaged in numerous informal talks, set up participatory mapping exercises, and studied documents, artifacts, and rituals.

The interview data have been transcribed and translated when needed, feeding into an analytical and interpretative process that has been somewhat different across research projects, as a result of the specific research questions of each. For this chapter, we have dug into our primary data to unveil insights on mobility.

We value relationship building and ethical commitment and try to the extent possible to communicate our findings back to the mining communities. This research is also embedded in the work that is done by CEGEMI (*Centre d'Expertise en Gestion Minière*), a research center based at the Catholic University of Bukavu.

Moving through time

The DRC has been highly dependent on mineral extraction since the colonial period. Today, its main exports are cobalt and copper, but it also exports tin, coltan, gold, and diamonds. While during the last decade there has been a revival of industrial large-scale mining, hundreds of thousands of people in the country are still dependent on artisanal and small-scale mining. Both industrial and artisanal mining have historically provoked massive movements of people: either through company recruitment strategies, or through rural poor people leaving their villages to work in the mines. This has had significant impacts on the ways in which ASGM and local economies tend to be organized. Furthermore, locals are often the landlords or ancestral heirs to the locations of ASGM sties. This ownership often implies several rules, such as cultural rules, used to control land and activities in their sites. In turn, this may lead to conflicts surrounding the question of autochthony (Dunia, Bikubanya, and Marijsse 2021).

While gold was extracted and used to make jewelry and royal attributes in pre-colonial Congo, it only acquired a real economic value with the arrival of colonial rule in the late nineteenth century, when Belgian prospectors began traversing Eastern DRC in search of easily accessible alluvial deposits. In 1923, MGL started mining gold in Mwenga territory (Geenen 2015). In Shabunda, *Compagnie Belge d'Entreprises Minières* (COBELMIN) and Kivumines were extracting cassiterite in Lulingu and Nzovu. In Misisi, no industrial mining company operated during the colonial period, although prospecting was undertaken. In Watsa territory, Australian prospectors found gold in 1903. In 1926, the Belgian government established the *Société des Mines d'Or de Kilo-Moto* (SOKIMO).

Local populations were displaced from their traditional forests and lands to make space for industrial mining. Some were forced to farm on *paysannats*, a system that privatized land and "rationalized" its use by requiring families to grow mainly cash crops, along with some food crops for their own consumption (Vinez 2017). Others were forced to work on the private and public road infrastructure that led to the mines (Biebuyck 1973). It is difficult to know how many people were forcibly displaced to work on the roads, and how many were attracted by the economic and social effects of the new roads. New villages were created by people who were forced to move. Other villages disappeared or were turned into agricultural plots that had to produce food for the mine workers. Rations were manually carried, mainly by adolescents and women, from the market outposts to the mines. These more-or-less forced displacements were done in cooperation with local chiefs.

To circumvent the issue of sparse local manpower and to alleviate pressure on the local population (birth rates were dwindling, but people also resisted forced labor), recruitment campaigns in neighboring countries and provinces brought in new workers (Geenen 2015). Up to this day, one street in Kamituga bears the name *chez les Barundi* (where the Barundi live), after the workers who came from Burundi. Once the workers settled in the mining towns, their movements were restricted, and their socio-cultural life was to a large extent controlled by the company (Heindryckx 2020; Mathys 2014). The company also tried to shape a new social identity. In Kamituga, a group of young men called themselves *les garcons de Kaga* (the boys from Kamituga) and cherished their identity as "sons" of the company. One of them expressed it as follows in a 2009 interview:

> You could not touch the company since it was like a parent to all of us. There was a degree of paternalism, yes. But it was difficult to step out of this logic. One could not imagine living without the company. People did not even leave Kamituga, because they didn't know where to go. Going to the village? That meant you were going to die because you would be bewitched.

However, Nancy Rose Hunt (2015) has recently argued that despite colonial restrictions on mobility, there is motion to be found in storytelling and narratives, which can both reveal and instigate mobility. In other words, the spatial and social reshaping by the colonizers ran into clear limits.

First, the displaced populations did not generally cut ties with their original living environment. Most continued to return to practice cultural ceremonies and other activities such as hunting and agriculture. Among the Balega, this gave rise to what is known as *matongo* in the local language (Kilega). *Matongo* is understood as the place of origin, the ancestral land, where people cohabitate with the elders who have passed away (Marijsse 2023). One of our interlocutors in Shabunda recalled: "The whites started to call people to come and settle along the road. It is because of this that each clan also has a *matongo* apart from being an occupant of a village far from his *matongo.*"

While the place of resettlement responded to the productive requests of the colonial authorities, the *matongo* responded to fundamental collective needs, closely tied to existence and life itself. Only in the *matongo* can ancestral spirits be brought in direct communication with humans. This is needed, for instance, upon the enthronement of a new clan chief, or for the management of intra-group disputes.

Second, although colonial companies attempted to completely reshape the identities of workers and their families, they did not fully succeed. Kamituga was formally an extra-customary locality, meaning that it was not ethnically homogeneous and customary law did not apply. In the more dispersed region of Shabunda, restrictive policies were fiercely contested, and cultural rites persisted. For example, repeated efforts to outlaw *mangene* (forest cabins used by hunters) did not prove successful. *Mangene* remained as outposts used by chiefs for cultural rituals, and by mine workers to flee the company.

Nevertheless, new social divisions were clearly created, not only between white "bosses" and Congolese laborers but also among the Congolese. Some were granted access to education and, consequently, attained social mobility as *évolués*, or those who have "evolved." Women were left to become "good housewives" and had very limited access to social mobility. Outside of the mining centers, most villagers continued to rely on farming, but because of the export-oriented *paysannats*, they became increasingly less self-reliant.

Around independence in 1960, people started panning for gold in rivers, and in Kamituga, a group of miners called *Ninja* began to descend at night in underground tunnels to "steal" gold from the company SOMINKI (*Société Minière et Industrielle du Kivu,* which succeeded MGL in 1976). In the 1970s and 1980s, the country plunged into a profound political, economic, and social crisis, owing to a number of factors, both external (commodity prices) and internal (detrimental nationalization policies, neopatrimonial politics) (Geenen 2015). Congolese were left to fend for themselves in the informal economy, which gave rise to a new national dictum: *débrouillez-vous*, or to fend for oneself. In 1982, President Mobutu liberalized gold and diamond mining. This gave a new impetus to the ASGM sector, pushing thousands of desperate villagers toward the mines (Dupriez 1987).

A new momentum developed in the period between 1996 and 2003, when coltan prices soared and ASGM became part of a regional war economy (Vlassenroot and Raeymaekers 2004). During this period, the mines not only offered much needed income opportunities but also, in many cases, relative security. Around a decade later, peak gold prices drew more and more people to the gold mines. In addition to rising prices, gold also became more attractive as it is easier to smuggle, while coltan and cassiterite came under increasing international scrutiny (Geenen 2015). New techniques and machinery, which will be discussed below, were introduced in Eastern DRC's ASGM sector from 2010 onward.

People on the move

A lot has been written about what motivates people to move into ASGM (among others, Banchirigah 2006; Hilson and Garforth 2012; Jønsson and Bryceson 2009; Maclin et al. 2017). But less attention has been paid to what they leave behind, what they take with them, and how they (re)build their selves at the site of destination or transit. In this section, we aim to provide some insights into the *connections* made in and through people's mobilities.

Miners we have studied not only move with the opening of economic opportunities, such as new gold discoveries or pits that go in and out of production. They also move with social relations, following colleagues and friends. For instance, dredge teams (known as *kazabuleurs*) generally move together as a group, along with their dredge on the rivers, although individual workers are also free to join other teams.

Often, artisanal miners were referred to as "itinerants" by our interlocutors, to signify that they are constantly on the move. And their movements are circular, as one interlocutor testified:

> This is the fifth time I have come back to work in Kachanga. At one point I went to Salamabila. I had returned here before going to the Nyange site where I had spent a few months. Now we are back in Kachanga. This is our life. We wander a lot.

In many sites, the landowners or pit managers are indigenous. They hire manual labor for doing tasks such as drilling, diving, and tunnel construction. Some of these specialized workers are foreigners who travel with their skills and particular techniques. In the region of Shabunda, *kazabuleurs* arrived around 2010. They traveled from western provinces of the DRC, such as Kasaï, to the Angolan frontier (during the civil war), to Kisangani, and then to Shabunda (Bumba Kamudiongo 2016; De Boeck 2001; Marijsse 2023; Omasombo 2001). They use rudimentary diving gear (such as pumps and air compressors) and skills previously used for diamond extraction (Geenen and Marijsse 2020; Marijsse 2023).

In Moku, technicians come from Uganda to maintain excavators and ball mills. In Misisi, Tanzanian technicians have introduced cyanidation as a technique to recover gold from tailings. These specialized workers often transfer their knowledge and skills. As an interlocutor explained, he had learned about the use of *testeurs* (a metal detector) from a Burundian miner (see also Dessertine 2016 for a West African example). Afterward, he went to different mining sites offering this service. In other cases, the specialized workers acquired their skills while working for, or around, industrial companies. For instance, there is one group of mercury workers who originated from the Twangiza area in Sud-Kivu, where they learned their skills from industrial miners. At present, their techniques are in high demand because they can handle large volumes of waste sand. Several timber workers we encountered had previously worked for the mining company SOMINKI and were now making the wooden structures to support underground tunnels.

This clearly testifies to the connection and interplay between industrial mining and ASGM. Such connections can also be seen through the shaping of miners' occupational identities. Just like colonial companies tried to shape a new identity – the "modern," industrious worker – the professional identity of ASGM miners is continuously taking shape. This identity is characterized by a sense of freedom, self-reliance, and independence from the state (Bryceson and Geenen 2016; Verweijen, Geenen, and Bashizi 2022). Some people have even maintained multiple identities, for instance, the *Ninja* who literally worked for the company SOMINKI during the day, and for themselves at night.

Technology on the move

In the past decade, machines and new techniques have gradually spread across Eastern DRC's ASGM sites. Table 5.1 provides a non-exhaustive overview of the main machines and techniques that we have identified.

From this table, we, first of all, observe that these machines and components are made in or transit via neighboring countries and enter DRC through bordering cities such as Baraka, Uvira, Kalemie, Bukavu, Goma, Bunia, and Aru (Lambertz 2021; Marijsse and Munga 2022). From there, they are transported to the mines not only by truck where possible but also by small cargo plane, improvised rafts, small boats, motorbikes or modified bikes (*kinga*), or manually on the backs of transporters. The choice of the means of transport, the time needed, and the cost of transport are tied to seasonal changes. During the rainy season, prices surge and the demand for manual carrying increases where trucks and motorcycles reach their limit. Also, waterways become riskier to navigate. The volatile security context obviously also renders some mining sites difficult to access.

Second, the pace and the extent of the spread of technologies clearly differ. As Dunia Kabunga and Geenen (2022, 5) observed, "The most widespread technologies are those that require relatively limited investment and can easily be transported." Large machines such as excavators and tractors are not only difficult to transport to remote sites but are also expensive (Bikubanya and Radley 2022). In some cases, large machines can be rented from industrial companies. We have found this to be the case in Babarau, where a subcontractor to Kibali Gold Mines rented an excavator to a group of ASGM miners. In Maï-Tongo, we witnessed how an excavator had dug a pit of 50 meters deep in less than a day, while a team of ten manual laborers would have needed at least two weeks to do so.

Table 5.1 Mobility of machines and techniques

Machines and techniques	*Origin or transit point*	*Use and spread*
Metal detectors (*testeurs*)	Tanzania; Burundi; Uganda	Not widespread. Owned by Congolese and used by Congolese technicians. Can prospect for gold at depths of 10–30 m.
Ball mills	Tanzania	Widespread. Owners were previously from neighboring countries, now Congolese. Workers are generally migrants.
Pans and *biporos* (banana sheaths)	Disseminated with the colonial companies, in Kamituga and other sites	Widespread. Used by indigenous workers.
Mercury	Probably spread with the colonial companies from Luhwinja or Burhinyi, in Mwenga territory in Sud-Kivu	Widespread. Introduced by migrants (for instance, in Kamituga and Shabunda, they are from Luhwinja and Burhinyi; in Moku and Durba, they are from Bunia and Aru).
Heating furnaces	Tanzania; Uganda	Widespread. Introduced by migrants.
Melting furnaces	Tanzania; Uganda	Not widespread (mostly Misisi). Introduced by migrant workers (mostly from Tanzania).
Cyanidation	Tanzania	Not widespread (one in Kamituga, three in Misisi and three in Baraka). Introduced by migrants from Tanzania.
Trucks (tippers)	Tanzania; Burundi; Uganda	Not widespread (mostly Misisi). Operated by indigenous workers.
Tractors	Tanzania; Burundi; Uganda	Not widespread (mostly Misisi, Kamituga and to a lesser extent Moku). Operated by Congolese, mostly trained in large-scale industrial mining.
Excavators	Tanzania; Uganda; Burundi	Not widespread (mostly Misisi and Moku). Operated by migrants.
Jackhammers	Tanzania; Burundi; Uganda	Widespread. Initially operated by migrants, now also Congolese workers.
Motor pumps	Tanzania; Uganda	Widespread. Operated by both migrants and Congolese.
Air compressors	Tanzania; Burundi; Uganda	Widespread. Operated by Congolese.
Explosives	Tanzania; Burundi; Uganda	Widespread. Operated by migrants (technicians from Burundi, Tanzania, Uganda) and Congolese (often trained in large-scale industrial mining).
Dredges	Tanzania; Uganda	Not widespread (Misisi on the Kimbi River, Durba on Kibali River and Shabunda on Ulindi River). Operated by migrants.
Robotic dredgers	Tanzania; Uganda	Not widespread. (Shabunda and Misisi). Operated by migrants.

Source: Adapted from Dunia Kabunga and Geenen (2022).

In some cases, local engineers also start to manufacture the machines or their components. In doing so, they make use of both the expertise and the material that has been left behind by the (post)colonial industrial companies. This is the case for mechanized ball mills, locally called *concasseurs* (Marijsse and Munga 2022). The same is true for some dredge parts, which are often locally "pirated" by welders from leftover iron scrap retrieved from SOMINKI's former workshops. In Kachanga and Babarau, we witnessed the use of explosives (to blast underground hard rocks) that came from industrial companies – either postcolonial, or the ones currently operating.

However, the machines used in large-scale industrial mining require specific expertise, and spare parts are often not on hand. A manager of a cooperative in Shabunda recounted that they had transported a Komatsu excavator from the port of Baraka through the forest region. They manually drove it, because they could not pay for a truck, and the road proved to be impassable with a truck wide enough to load an excavator. They left the port in December 2019 and reached the mining site one and a half years later, in May 2021. But, when its caterpillar track gave in, the excavator remained stuck at the site. It has proved impossible to repair, as an interviewee recounted:

> We tried to force the excavator to roll. Because of that, it broke down.... It's very complicated [to repair it here]. There is a part on the ring gear that weighs more than 15 tons. There is no way to get the crane here that can help us to lift and replace this ring gear.

Matter on the move

We started this chapter by observing that miners are "hunting the gold." In this section, we make this more concrete by looking at prospecting techniques, processing of tailings, and the disappearance of gold. In doing so, we show how the resource assemblage is put to work, and how it involves human agency as well as the agency of matter and technology.

First, we look at prospecting techniques. In underground mining, gold veins are made visible through socio-technical practices called *meta*. *Meta* draws on hearsay, manual sampling, and "reading" the environment. Miners may receive information from other miners, or from family members who used to work for (post)colonial mining companies. This information sometimes involves wild tales about striking it rich or suspicions about hidden gold left behind by colonizers (Morris 2008). When proceeding underground, mine workers also take samples, which are analyzed above ground. Finally, *meta* involves being attuned to the environment, reading the landscape for traces of colonial presence, such as prospection pits, tailings, and the presence of non-native trees that were planted to provide wood for tunnel construction (Geenen 2015). Here, damaged soils, vegetation, hearsay, and things that feel out of place or uncanny fuel prospection practices. Recently, however, the use of *testeurs* or metal detectors has changed these prospection practices. This not only has allowed prospection to become more technology-driven; it also has given rise to a new group of workers. Similar to what Calkins (2016) describes in the case of Sudan, these workers are typically young, skilled, and mobile.

In the case of underwater dredging, *testeurs* cannot be used. Here, prospection is done manually, making use of a metal funnel to scoop up three to four samples (Marijsse 2023). On the surface, the sand is panned and evaluated to further direct the work of the divers. When different dredgers work in each other's vicinity, the workers who check for the presence of gold in the samples are allowed to check the sluice boxes of other dredges for gold specks. Sometimes, divers merely follow hearsay, pressured by the necessity of arriving in time. Very often, the challenge of transport requires dredgers to make fast decisions when minerals are found in high quantities somewhere. There is simply no time to lose. Then the team will decide either to disassemble the dredge or transport it elsewhere, or to maneuver it on the river using motorized boats.

Second, material that was previously considered to be waste is made profitable again, thanks to the introduction of machines and new techniques (see also Lanzano and Di Balme 2021; Pijpers et al. 2021; Voyles 2015). Mechanized ball mills, for instance, can process larger volumes of stones and soil, so that even the processing of lower-grade deposits becomes profitable (Bikubanya and Radley 2022). Cyanidation allows for a higher recovery rate than manual techniques using gravitation, or mercury amalgamation (Nkuba, Muhanzi, and Zahinda 2022). The process has recently been introduced in Misisi and in Kamituga, with technology and experts from Tanzania.

Regardless of the specific technologies used, there is an almost infinite level of human ingenuity demonstrated in the process of extracting, working, and reworking matter (Dunia Kabunga and Geenen 2022). Around ball mills, for instance, female stone pickers search for remaining mineralized stones. Others scrape the inside of the ball mill, calling it *kokora*, after the word for burnt residue in a saucepan in Kiswahili. In Shabunda, a general term is *makalo*, after the Kilega word for the remains of pressed palm nuts. In Misisi, the workers collecting leftovers are called *bindistes*, from the Kibembe word *bindi* meaning waste. Here, some workers even burn the bags in which stones and sand have been transported. "After burning, they obtain a blackish material. According to some gatherers it takes at least 40 bags to fill a spade with the blackish material, in which you can find a minuscule quantity of gold, if you are lucky" (Dunia Kabunga and Geenen 2022, 5). As such, *matter* constantly transforms and moves in and out of the resource-state.

Third, gold can also (be made to) disappear. In Babarau, we were told about the phenomenon of "owls" (*hibouneurs*) who enter the pits at night to steal bags of gold-containing soil, and who often block the pit entrance with stones or sand when they come out. According to our interlocutors, this occurred more frequently in pits owned by non-natives. In this sense, gold disappears during the night, very much like what happened during the 1970s and 1980s in SOMINKI's underground tunnels. At the time, SOMINKI put in place stringent security measures: the tunnel gates were guarded, the factory was walled, a double door protected the amalgamation room, a metal detector was installed, and all three working shifts were supervised by an "expat" (Geenen 2015), although this did not stop the infiltrations. Similarly, pit managers in Babarau have stationed night guards in an attempt to prevent losses.

In Shabunda, we were told about cases of gold disappearance that were the result of witchcraft. Some of the victims of these practices have sought help from traditional healers in the area. A farmer and his sister testified:

> We were at this old man's house. K was complaining that when she has money, she feels it mysteriously disappears. This old man had helped her. He is very good at stories of luck, domination, and disappearance. And since she had been to see this old man, business is going quite well.

In the town of Misisi and the surrounding sites, itinerant traders from Burundi or other Congolese provinces, such as Tanganyika and Maniema, sell amulets protecting against evil spirits. In Babarau and Durba, amulets are sold by traders from Central African Republic, Uganda, and South Sudan. All this testifies to the idea that gold is not fixed. It comes and it goes, not only as a consequence of geological forces but of socio-cultural interventions too.

Conclusion

When ASGM miners travel, they never travel alone. Materials, capital, technologies, skills, and know-how travel with them. These human and non-human elements gather into a resource assemblage. The assemblage works to prospect for gold and to render extraction possible and profitable. As such, extraction not only requires motion, as in removing material from the underground. It also relies on motion; on the mobility of humans, technology/techniques, and materials.

Historically speaking, this chapter has discussed a number of critical junctures – colonization, liberalization, mechanization – as places where *connections* have been forged. Processes occurring during these junctures have stimulated labor migration from neighboring countries to Eastern DRC as well as from rural to rapidly urbanizing areas, but they have not erased previous identities, cultures, techniques, and know-how. We have also shown the transportation of technologies and machines, their connections to regional trade patterns, the obstacles they encounter *en route* to the mines, and the ingenuity of local engineers in adapting the machines to their environment. Finally, we have highlighted movement in matter; that is, the rocks and sand being removed from the underground, and the gold contained in these. We have discussed prospecting techniques making gold visible, machines making waste valuable again, and nightly thieves and traditional healers making gold disappear or reappear.

In brief, gold is located, suspected, and detected in different forms and settings and extracted through continuously changing means that imply global and cross-regional flows. This requires us to *move ourselves* toward a more-than-human analytical frame (Kohn 2013) and to locate agency in between relationalities that imply the occult, geological specificity, migration, and technology (D'Angelo 2019; Nystrom 2014; Rolston 2013).

This observation is paramount to the domain of anthropology of mining and mobility because it locates ASGM at the intersection of different forms of knowledge and technology transfer and embedded in a multilayered historical context. In sum, what our chapter has foregrounded through the life stories of our itinerants is that ASGM indeed is rife with contingent encounters, seasonality, risk, luck, trial and error, occult interventions, and often sheer randomness (D'Angelo 2015). And yet, it cannot be reduced to this. Technologies that navigate underwater, underground, and above the surface also require skill, apprenticeship, know-how, and the transport and modification of this hardware (Lambertz 2021; Marijsse and Munga 2022). They all join forces in ASGM as a moving project.

References

Banchirigah, Sadia M. 2006. "How Have Reforms Fueled the Expansion of Artisanal Mining? Evidence from Sub-Saharan Africa." *Resources Policy* 31 (3): 165–171.

Banchirigah, Sadia M., and Gavin Hilson. 2010. "De-Agrarianization, Re-Agrarianization and Local Economic Development: Re-Orienting Livelihoods in African Artisanal Mining Communities." *Policy Science* 43: 157–180.

Biebuyck, Daniel. 1973. *Lega Culture: Art, Initiation, and Moral Philosophy Among a Central African People.* Berkeley: University of California Press.

Bikubanya, Divin-Luc, and Ben Radley. 2022. "Productivity and Profitability: Investigating the Economic Impact of Gold Mining Mechanisation in Kamituga, DR Congo." *The Extractive Industries and Society* 12: 101162.

Bolay, Matthieu. 2016. "'Il Faut êTre Là Où L'or Sort!': De L'itinérance Temporaire au Maintien d'un Mode de Vie Mobile Chez les Orpailleurs de Haute Guinée." *Vienna Journal of African Studies* 30 (16): 111–135.

Bridge, Gavin. 2009. "Material Worlds: Natural Resources, Resource Geography and the Material Economy." *Geography Compass* 3 (3): 1217–1244.

Bryceson, Deborah F., and Sara Geenen. 2016. "Artisanal Frontier Mining of Gold in Africa: Labour Transformation in Tanzania and the Democratic Republic of Congo." *African Affairs* 115 (459): 296–317.

Bumba Kamudiongo, G. 2016. "Dancing to the Drums (Ngoma) Of the Youth. At the Crossroads of Shifting Cultural Grounds: The Bana Lunda Youth as Agents of Transformation." PhD diss., KU Leuven.

Calkins, Sandra. 2016. "How 'Clean Gold' Came to Matter: Metal Detectors, Infrastructure, and Valuation." *HAU: Journal of Ethnographic Theory* 6 (2): 173–195.

Cuvelier, Jeroen. 2011. "Men, Mines and Masculinities: The Lives and Practices of Artisanal Miners in Lwambo (Katanga Province, Drcongo)." PhD diss, KU Leuven.

Dahinden, Janine. 2010. "Cabaret Dancers: 'Settle Down in Order to Stay Mobile?' Bridging Theoretical Orientations within Transnational Migration Studies." *Social Politics* 17 (3): 323–348.

D'Angelo, Lorenzo. 2014. "Who Owns the Diamonds? The Occult Economy of Diamond Mining in Sierra Leone." *Africa* 84 (2): 269–293.

D'Angelo, Lorenzo. 2015. "Diamond Mining Is a Chain. Luck, Blessing and Gambling in Sierra Leone's Artisanal Mines." *Critical African Studies* 7 (3): 243–261.

D'Angelo, Lorenzo. 2019. "God's Gifts: Destiny, Poverty, and Temporality in the Mines of Sierra Leone." *Africa Spectrum* 54 (1): 44–60.

D'Angelo Lorenzo. 2022. "Technology." In *The Anthropology of Resource Extraction,* edited by Lorenzo D'Angelo and Robert J. Pijpers, 167–184. London: Routledge.

De Boeck, Filip. 2001. "Garimpeiro Worlds: Digging, Dying & 'Hunting' for Diamonds in Angola." *Review of African Political Economy* 28 (90): 549–562.

Dessertine, Anna. 2016. "From Pickaxes to Metal Detectors: Gold Mining Mobility and Space in Upper Guinea, Guinea Conakry." *The Extractive Industries and Society* 3: 435–441.

Dibwe dia Mwembu, Donatien. 2001. *Bana Shaba abandonnés par leur père: structures de l'autorité et histoire sociale de la famille ouvrière au Katanga, 1910–1997* [Bana Shaba Abandoned by Their Father: Structures of Authority and Social History of the Working Family in Katanga, 1910–199]. Paris: l'Harmattan.

Dunia Kabunga, Philippe, Divin-Luc Bikubanya, and Simon Marijsse. 2021. "Croyances Culturelles ET Effectivité Sur L'Exploitation Minière Artisanale en Rd Congo: Entre Contestation ET Négociation." In *Conjonctures de l'Afrique centrale 97*, edited by Ndayiragije, Réginas, Alidou Sahawal, An Ansoms, and Sara Geenen (dir.), 236–261. Tervuren, Paris: Musée Royal de l'Afrique centrale et l'Harmattan.

Dunia Kabunga, Pjilippe, and Sara Geenen. 2022. "Work Regimes and Technology-Driven Transformations in Dr Congo's Artisanal and Small-Scale Gold Mines." *The Extractive Industries and Society* 12: 101142.

Dupriez, Hugues. 1987. *Bushi: L'Asphyxie D'Un People* [Bushi: The Suffocation of a People]. Bukavu: ADI-Kivu.

Fetter, Bruce. 1973. "L'Union Minière du Haut-Katanga, 1920–1940: La Naissance D'Une Sous-Culture Totalitaire." *Cahiers du CEDAF* 6: 1–40.

Geenen, Sara. 2015. *African Artisanal Mining from the inside Out: Access, Norms and Power in Congo's Gold Sector.* London: Routledge.

Geenen, Sara, and Simon Marijsse. 2020. "The Democratic Republic of Congo: From Stones in the River to Diving for Dollars." In *Global Gold Production Touching Ground: Expansion, Informalization, and Technological Innovation*, edited by Verbrugge, Boris and Sara Geenen, 263–281. London: Palgrave MacMillan.

Geschiere, Peter. 1997. *The Modernity of Witchcraft: Politics and the Occult in Postcolonial Africa*. Virginia: University of Virginia Press.

Grätz, Tilo. 2003. "Gold-Mining and Risk Management: A Case Study from Northern Benin." *Ethnos* 68 (2): 192–208.

Harries, Patrick. 1994. *Work, Culture, and Identity: Migrant Labourers in Mozambique and South Africa, c. 1860–1910.* London: James Currey.

Heindryckx, Laurence. 2020. "Governing Economic Interests: Interwar Road Construction in Belgian Congo." *Belgeo. Revue Belge de Géographie.* doi: https://doi.org/10.4000/belgeo.43202.

Hilson, Gavin, and Chris Garforth. 2012. "Everyone Now Is Concentrating on the Mining: Drivers and Implications of Rural Economic Transition in the Eastern Region of Ghana." *Journal of Development Studies* 49 (3): 348–364.

Hilson, Gavin, Abigail Hilson, Roy Maconachie, James McQuilken, and Halima Goumandakoyea. 2017. "Artisanal and Small-Scale Mining (Asm) In Sub-Saharan Africa: Reconceptualizing Formalization and 'Illegal' Activity." *Geoforum* 83: 80–90.

Hilson, Gavin, and Roy Maconachie. 2020. "Entrepreneurship and Innovation in Africa's Artisanal and Small-Scale Mining Sector: Developments and Trajectories." *Journal of Rural Studies* 78: 149–162.

Hunt, Nancy R. 2015. *A Nervous State*. Durham: Duke University Press.

Jeeves, Alan H. 1985. *Migrant Labour in South Africa's Mining Economy: The Struggle for the Gold Mines' Labour Supply, 1890–1920.* Montreal: McGill-Queen's University Press.

Jønsson, Jesper B., and Deborah F. Bryceson. 2009. "Rushing for Gold: Mobility and Small-Scale Mining in East Africa." *Development and Change* 40 (2): 249–279.

Kohn, Eduardo. 2013. *How Forests Think: Toward an Anthropology Beyond the Human.* Berkeley: University of California Press.

Lambertz, Peter. 2021. "Longola Marche Arrière! Chinese Diesel Engines on Congo's Inland Waterways." *Critical African Studies* 15 (1): 15–34.

Lanzano, Cristiano. 2020. "Guinea Conakry and Burkina Faso: Innovations at the Periphery." In *Global Gold Production Touching Ground: Expansion, Informalization, and Technological Innovation*, edited by Verbrugge, Boris and Sara Geenen, 245–262. London: Palgrave MacMillan.

Lanzano, Cristiano, and Luigi Arnaldi di Balme. 2021. "Who Owns the Mud? Valuable Leftovers, Sociotechnical Innovation and Changing Relations of Production in Artisanal Gold Mining (Burkina Faso)." *Journal of Agrarian Change* 21 (3): 433–458.

Li, Tania M. 2014. "What Is Land? Assembling a Resource for Global Investment." *Transactions of the Institute of British Geographers* 39 (4): 589–602.

Maclin, Beth, Jocelyn Kelly, Rachel Perks, Patrick Vinck, and Phuong Pham. 2017. "Moving to the Mines: Motivations of Men and Women for Migration to Artisanal and Small-Scale Mining Sites in Eastern Democratic Republic of the Congo." *Resources Policy* 51: 115–122.

Mainwaring, Ċetta, and Noelle Bridgen. 2016. "Beyond the Border: Clandestine Migration Journeys." *Geopolitics* 21 (2): 243–262.

Mairie de Kamituga. 2022. *Rapport de Statistiques de la Population de la Ville de Kamituga, Trimestre 1/2022* [Kamituga City Population Statistics Report, Quarter 1/2022]. Kamituga. Unpublished.

Marijsse, Simon. 2023. "Ambient Technologies. An Ethnography of Extractive Pathways in Eastern Dr Congo." PhD diss., Universiteit Antwerpen & KU Leuven.

Marijsse, Simon, and Thierry Munga Mwisha. 2022. "Taming Air and Water: The Fight against Shimoke in Artisanal and Small-Scale Gold Mining in South Kivu." *The Extractive Industries and Society* 12: 101168.

Massaro, Luciana, and Marjo de Theije. 2018. "Understanding Small-Scale Gold Mining Practices: An Anthropological Study on Technological Innovation in the Vale Do Rio Peixoto (Mato Grosso, Brazil)." *Journal of Cleaner Production* 204: 618–635.

Mathys, Gillian. 2014. "People on the Move: Frontiers, Borders, Mobility and History in the Lake Kivu Region 19th-20 century." PhD diss., University of Ghent.

Mkodzongi, Grasian, and Samuel Spiegel. 2020. "Mobility, Temporary Migration and Changing Livelihoods in Zimbabwe's Artisanal Mining Sector." *The Extractive Industries and Society* 7 (3): 994–1001.

Moodie, T. Dunbar. 1994. *Going for Gold. Men, Mines and Migration.* Berkeley: University of California Press.

Mol, Annemarie. 2002. *The Body Multiple: Ontology in Medical Practice.* Durham, NC: Duke University Press.

Morris, Rosalind C. 2008. "The Miner's Ear." *Transition* 98: 96–115.

Nkuba, Bossissi, Aganze G. Muhanzi, and Zahinda Mugisho Franck. 2022. "How Do Technological Changes in Artisanal and Small-Scale Gold Mining Affect the Environment and Communities' Health?" *The Extractive Industries and Society* 12: 101178.

Nystrom, Eric C. 2014. *Seeing Underground: Maps, Models, and Mining Engineering in America.* Reno: University of Nevada Press.

Omasombo, Jean. 2001. "Les Diamants de Kisangani: De Nouveaux Seigneurs SE Taillent Des Fiefs Sur Le Modèle de L'éTat Zaïrois de Mobutu." In *Chasse AU Diamant AU Congo-Zaïre*, edited by Monnier, Laurent, Bogumil Jewsiewicki, and Gautier de Villers, 79–126. Tervuren: Institut Africain-CEDAF.

Peša, Iva, and Benoît Henriet. 2021. "Beyond Paternalism: Pluralising Copperbelt Histories." In *Across the Copperbelt: Urban & Social Change in Central Africa's Borderland Communities*, edited by Larmer, Miles, Rachel Taylor, Enid Guene, Iva Peša, and Benoît Henriet, 27–52. Woodbridge: Boydell & Brewer.

Pfaffenberger, Bryan. 1998. "Mining Communities, Chaînes Opératoires and Sociotechnical Systems." In *Social Approaches to an Industrial Past: The Archaeology and Anthropology of Mining,* edited by A. Bernard Knapp, Vincent C. Pigott, and Eugenia W. Herbert, 291–300. London: Routledge.

Pijpers, Robert, Esther van de Camp, Eleanor Fisher, Luciana Massaro, Jorge Calvimontes, Lorenzo D'Angelo, and Cristiano Lanzano. 2021. "Mining 'Waste'." *Etnofoor* 33 (2): 13–40.

Poste du secteur de Ngandja. 2021. *Rapport annuel*. Unpublished document.

Read, Margaret. 1942. "Migrant Labour in Africa and Its Effects on Tribal Life." *International Labor Review* 45 (6): 605–631.

Richards, Audrey. 1932. "Anthropological Problems in North-Eastern Rhodesia." *Africa* 5 (2): 121–144.

Rolston, Jessica S. 2013. "The Politics of Pits and the Materiality of Mine Labor: Making Natural Resources in the American West." *American Anthropologist* 115 (4): 582–594.

Rubbers, Benjamin. 2019. "Mining Towns, Enclaves and Spaces: A Genealogy of Worker Camps in the Congolese Copperbelt." *Geoforum* 98: 88–96.

Schapendonk, Joris, Ilse Van Liempt, Inga Schwartz, and Griet Steel. 2018. "Re-Routing Migration Geographies: Migrants, Trajectories and Mobility Regimes." *Geoforum* 166: 211–216.

Schapera, Isaac. 1947. *Migrant Labour and Tribal Life: A Study of Conditions in the Bechuana Protectorate*. Oxford: Oxford University Press.

Tarrius, Alain. 2002. *La Mondialisation Par Le Bas. Les Nouveaux Nomades de L'éConomie Souterraine* [Globalization From Below. The New Nomads of the Underground Economy]. Paris: Balland.

Territoire de Shabunda. 2019. *Rapport sur l'exploitation minière artisanale en territoire de Shabunda* [Report on Artisanal Mining in Shabunda Territory]. Shabunda. Unpublished.

Valdivia, Gabriela, Matthew Himley, and Elizabeth Havice. 2021. "Critical Resource Geography: An Introduction." In *The Routledge Handbook of Critical Resource Geography*, edited by Himley, Matthew, Elizabeth Havice, and Gabriela Valdivia, 1–20. New York: Routledge.

Van der Velde, Martin, and Ton van Naerssen. 2011. "People, Borders, Trajectories: An Approach to Cross-Border Mobility and Immobility in and to the European Union." *Area* 43 (2): 218–224.

Van Onselen, Charles. 2001. *New Babylon, New Nineveh: Everyday Life on the Witwatersrand*. Jeppestown: Jonathan Ball.

Verbrugge, Boris, Cristiano Lanzano, and Matthew Libassi. 2021. "The Cyanide Revolution: Efficiency Gains and Exclusion in Artisanal and Small-Scale Gold Mining." *Geoforum* 126: 267–276.

Verweijen, Judith, Sara Geenen, and Anuarite Bashizi. 2022. "Articulating Sedimented Subjectivities: Extractive Subject Formation in Eastern DRC." *Geoforum*. doi: 10.1016/j.geoforum.2022.10.009.

Vinez, Margaux. 2017. *Division of the Commons and Access to Land on the Frontier: Lessons from the Colonial Legacy in the Democratic Republic of Congo*. Washington, DC: The World Bank.

Vlassenroot, Koen, and Timothy Raeymaekers. 2004. *Conflict and Social Transformation in Eastern DR Congo*. Ghent: Academia Press.

Voyles, Traci B. 2015. *Wastelanding: Legacies of Uranium Mining in Navajo Country*. Minneapolis: University of Minnesota Press.

Walsh, Andrew. 2003. "'Hot Money' and Daring Consumption in a Northern Malagasy Sapphire-Mining Town." *American Ethnologist* 30 (2): 290–305.

Werthmann, Katja. 2003. "Cowries, Gold and 'Bitter Money'. Gold-Mining and Notions of Ill-Gotten Wealth in Burkina Faso." *Paideuma* 49: 105–124.

6 Making mining localities

Trajectories and stories of mining and mobility in Zambia

Patience Mususa and Iva Peša

Introduction

Ever since the Zambian Copperbelt, a stretch of richly mineralized land from Chililabombwe in the northwest to Ndola in the southeast, started urbanizing in the 1930s, mining and regional migration have been studied in tandem (e.g., Epstein 1958; Mitchell 1966; Heisler 1974). Rhodes Livingstone Institute anthropologists in the 1940s and 1950s set the tone with their inquiries into the relationship between migration, urbanization, and social change (Wilson 1941; Epstein 1958; Gluckman 1960; Schumaker 2001). In the decades since, debates have ranged from whether labor migrants would turn into "permanent urbanites" (Ferguson 1990; MacMillan 1993) and how kinship ties would be reconfigured in urban areas (Ferguson 1999; Larmer 2021), to whether a copper mining slump and economic recession had caused "counter-urbanization" in the 1990s (Potts 2005; Mususa 2021). These works are all premised on the assumption that copper mining itself is the inherent driving force of migratory movements and urbanization (Bryceson et al. 2022). The post-2000 opening of new copper mines in Zambia's North Western province has revived many of these debates about the relationship between mining, migration, and urban development (Negi 2014; Kesselring 2018).

In this chapter, we investigate the complex relationship between mining and mobility in Zambia to better understand how mining localities have been made and transformed over time. We aim to bring survey results, life histories, and the rich scholarship on regional mining localities into conversation to generate new insights into the dynamics that underpin mobility in Zambian mining hubs. We focus on three mining localities in particular: Mufulira and Luanshya, two mining districts in the Copperbelt province that were established at the beginning of the twentieth century; and Kalumbila, a newer mining district in the North Western province. These localities were selected because they enable us to compare places where growth has tapered off with a place that is in full development. We present our analysis in four parts. After introducing our three case study localities and explaining the methodology, we outline possible drivers and motives for mobility in mining localities. Thereafter, we discuss the outcomes of population survey research in Kananga, a settlement within Kalumbila district, and then we zoom in on individual life history trajectories from Mufulira, Luanshya, and Kalumbila,

DOI: 10.4324/9781003313236-8

as these provide us with a richer picture of mobility in Zambian mining localities. Applying this multi-methods approach and contextualizing our analysis within the rich scholarship on regional mining localities in Zambia, we demonstrate that mining-related mobility in the country has been influenced by a range of factors apart from the availability of mining employment. Other important drivers and enablers include socio-economic conditions in "source" or "home" communities; opportunities in agricultural production, forestry, and trade in – and adjacent to – mining areas; and the connections created through social ties and networks (Bryceson and MacKinnon 2012; Udelsmann Rodrigues and Tavares 2012; Udelsmann Rodrigues et al. 2021).

Our chapter goes beyond a narrow focus on migration to investigate patterns of mobility more broadly, including movements that may be over shorter rather than longer distances, and temporary, itinerant, or circular, rather than permanent. We take our departure point from the scholarship on place-making (Castillo and Brereton 2018; Luning 2018) that centers on understanding the transformation of places, not as cartographic space, but as linked to the conglomeration of multiple lived itineraries. Localities are, as Doreen Massey (1993, 145) has argued, the "relations of interdependence" to other places that in turn produce the particularities of place. Large-scale mining activities, such as those that dominate Zambia's economy, link extractive locales to the wider world. Yet the dynamics of these locales are not everywhere the same. Mining activity, through boom and bust, entangles with other livelihood activities to co-shape new understandings of place and new forms of mobility.

By unpacking the complex patterns of human mobility through which mining localities are made, we also challenge the concept of mining hubs as "enclaves" (Ferguson 2005; Auty 2006; Rubbers 2019). Instead, we emphasize the multiple linkages that mining hubs forge and maintain with their expansive hinterlands (Larmer et al. 2021; Peša 2022). Moreover, we underline that human mobility in Zambia has been influenced by a range of factors, only one of which is mining. Deborah Bryceson et al. (2022) likewise argue that while mining has been an important driver of urbanization processes and mobility in extractive locales in Africa, mining also interacts with other longer established economic activities such as agriculture.

Setting the scene: methodology and case study selection

This chapter draws on both quantitative and qualitative data. The qualitative data comprises life histories recorded through interviews with long-term Copperbelt residents living in Mufulira and Luanshya, as well as with inhabitants of Kalumbila. These interviews were conducted by the authors, both of whom have carried out extensive research in various Zambian mining localities. The presented life histories, selected from a sample of over 100 interviews, are illustrative of broader patterns of mobility and complicate insights from the secondary literature. The quantitative data comes from two sources: the 2020 and 2022 national census, which provide statistical information on population trends in various mining districts; and a household survey, carried out in October 2021 in the settlement of Kananga, in Kalumbila district, a fast-developing region in Zambia's North Western province. The survey, covering 62 respondents, was carried out over a period of three days in October 2021 by a team of five survey administrators.[1] They conducted in-person interviews in five separate sections of Kananga and visited every other household where an informed respondent was available.[2] The survey included questions on migration histories; mobility patterns; relationships to other locations; work, housing, and neighborhood transformations over time; and projections of the future. It was undertaken as part of a much larger study looking at processes of urban consolidation across Angola, Democratic Republic of the Congo (DR Congo), and Zambia, with a particular focus on boomtowns, peripheral urbanization, and settlements arising from displacement.[3]

Overview of case study areas

Until the commodities boom of the early 2000s, mining in Zambia had largely been concentrated in the Copperbelt province, where industrial mining has been ongoing since the early twentieth century. Since the early 2000s, new mining activity is increasingly taking place in the North Western province, a region adjacent to the Copperbelt province. (Both neighbor the mineral-rich region of Haut Katanga in DR Congo.) Together, these areas present an interesting opportunity to look at both the *longue durée* of mining and mobility in Zambia as well as more contemporary experiences related to new mining areas. This is what motivates our selection of cases. Mufulira and Luanshya, established in the 1930s, are some of the oldest industrial mining districts in the Copperbelt province. Kalumbila is a new district, established in 2018, to support mining in the North Western province. Like in the older established Copperbelt province, its mining districts comprise various settlements and areas. There is what is often referred to as the mine township, which is the settlement built by the mining companies to house its workers. In addition to the mine townships, there are the municipal townships, which are settlements directly administered by the municipal authorities. There are also customary areas, settlements that fall under customary authority. These would include Kananga, in Kalumbila district, which used to be rural but may now be described as 'rurban' – i.e., rural, but with urban characteristics. Also integrated within these districts are forest areas and farming blocks. In the case of Kalumbila district, there is also a large refugee camp. Old and new mining activity in these areas did not occur on a tabula rasa. Nevertheless, as we show below, mining has had an influence on settlement and mobility.

In 2022, Zambia conducted a national census that confirmed the booming dynamics of the mining districts of Kalumbila and Solwezi in the North Western province (see Figure 6.1), which had already been observed by scholars of the region (Negi 2014). The North Western province had an average annual population growth rate of 4.8 percent between 2010 and 2022 (ZSA 2022, 12), a significant increase from an annual growth rate of 2.2 percent between 2000 and 2010 (CSO 2012, 3). Much of this change has been driven by the development of large-scale mines since 2005 in the Solwezi and Kalumbila districts, which between 2010 and 2022 had annual population growth rates of 8 percent and 8.1 percent, respectively (ZSA 2022, 25). This growth occurred in a region long known as predominantly rural, dominated by the agricultural sector. Mining in the North Western province is catalyzing urbanization, reflected in the near doubling of the urban percentage of the population in the province, from 22.6 percent in 2010 (CSO 2012, 41) to 41.7 percent in 2022 (ZSA 2022, 11). The survey conducted in Kananga, in Kalumbila district, presented in the next section, highlights some of the granular aspects of mobility connected to this growth.

In contrast, the Copperbelt province, where the districts of Mufulira and Luanshya are located, had in the period between 2010 and 2022 one of the lowest annual population growth rates in the country, at 2.8 percent (ZSA 2022, 12). In the same period, Luanshya had an annual growth rate of 2.6 percent, whereas that of Mufulira was only 1.7 percent (ZSA 2022, 22). While markedly lower than the annual growth rates of Kalumbila district, both Copperbelt districts have experienced a kind of resurgence from the 2000 to 2010 census period, when annual average population growth rates were 0.5 percent for Luanshya and 1.2 percent for Mufulira (CSO 2012, 15). This indicates a tentative recovery from the worst years of economic crisis that followed the re-privatization of the Zambia Consolidated Copper Mines (Mususa 2021).

Apart from population dynamics, how can these two districts be typified? For one, they are characteristically referred to as mining towns, even as the districts have a variety of settlement types and livelihood activities. Mufulira is a town built by mining. The first copper deposits were identified there in 1923, and the town gained municipal status in 1937. Population numbers rose from 50,000 in 1956 to 150,000 in 1980, 162,889 in 2010, and 200,182 in 2022. Mufulira has consistently been one of the biggest copper producers in Zambia's mining

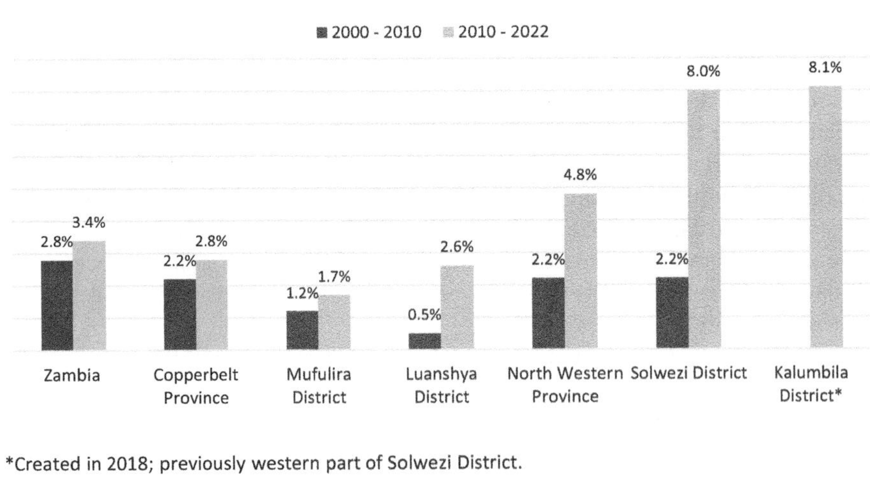

Figure 6.1 Average annual population growth rates (percent) in case study areas for the 2000–2010 and 2010–2022 periods. Data sources: CSO 2012 and ZSA 2022.

industry, although production was temporarily upset following the Mufulira mine disaster of 1970, when a massive tailings inrush inundated the mine and killed 89 workers. The development of deep-level mining in the 1980s restored production and Mufulira mine today boasts a shaft reaching almost two kilometers in depth. Throughout its history, Mufulira has attracted migrants from all parts of Zambia and from neighboring Congo. Besides mining, Mufulira is also a major trading center, particularly for goods from Congo and for dried fish from Luapula (Peša and Henriet 2021).

Luanshya's Roan Antelope copper mine was developed from 1926 onward. Copper deposits were located in a swampy area of the Luanshya River, necessitating draining to prevent high fatalities due to malaria (Schumaker, 2008). Luanshya was the epicenter of the region's first mining strike, in 1935, and trade unionism remained high since. Luanshya mine was nationalized in 1970 and reprivatized again in 1997. A major flood caused a stop in operations between 2001 and 2010, but production resumed after the mine was bought by a Chinese company. While copper production continues, many workers have been retrenched and the town is facing a crisis (Straube 2021). Connected by rail, road, and air to other Zambian urban areas, Luanshya has a variety of factories, which include copper wire production and metal fabrication. Over the years, Luanshya's population has risen steadily, from 96,000 in 1969 to 156,059 in 2010, and 211,996 in 2022.

Both Luanshya and Mufulira have significant farming activity that grew as mining livelihoods diminished (Smart 2015; Umar 2017), as well as a forest industry, that includes logging and charcoal production (Mutamba 2007).

Kalumbila district in the North Western province was established in 2018 to bring municipal administration close to the areas where the two large-scale mines of Lumwana and Kalumbila had been developed circa 2007 and 2010, respectively (see maps in Figure 6.2). As a result of all this mining activity, Kalumbila's population grew rapidly from 85,505 in 2010 to 177,067 in 2022 (ZSA 2022, 20). In addition to two large-scale mines, and their respective mine townships (Kalumbila and Lumwana), Kalumbila district also has a new municipal township area. It has several settlements that are growing, such as Manyama and Kananga (the latter of which this chapter includes survey data on). It also includes a large refugee settlement, Meheba, established to host Angolan refugees in the 1970s, and several chieftaincies, one of the largest being Chief Musele of the Lunda people, whose people had long settled in the region. Other than mining, agriculture, concentrated within and around the Meheba refugee camp, is important (Nyamazana et al. 2017). So too are forest-based livelihood activities such as beekeeping (Mickels-Kokwe 2006).

Even though the Sentinel copper mine in Kalumbila holds one of the world's largest remaining copper reserves, production only commenced in 2016, after First Quantum Minerals acquired the mine in 2010. Kalumbila is located over 300 km northwest of Kitwe, the main Copperbelt center, along a poorly maintained road. When the mine township was being developed, Zambia's erstwhile President, Edgar Lungu, touted Kalumbila mine township as a flagship project: "This will now become a new model town for all developing towns in Zambia"

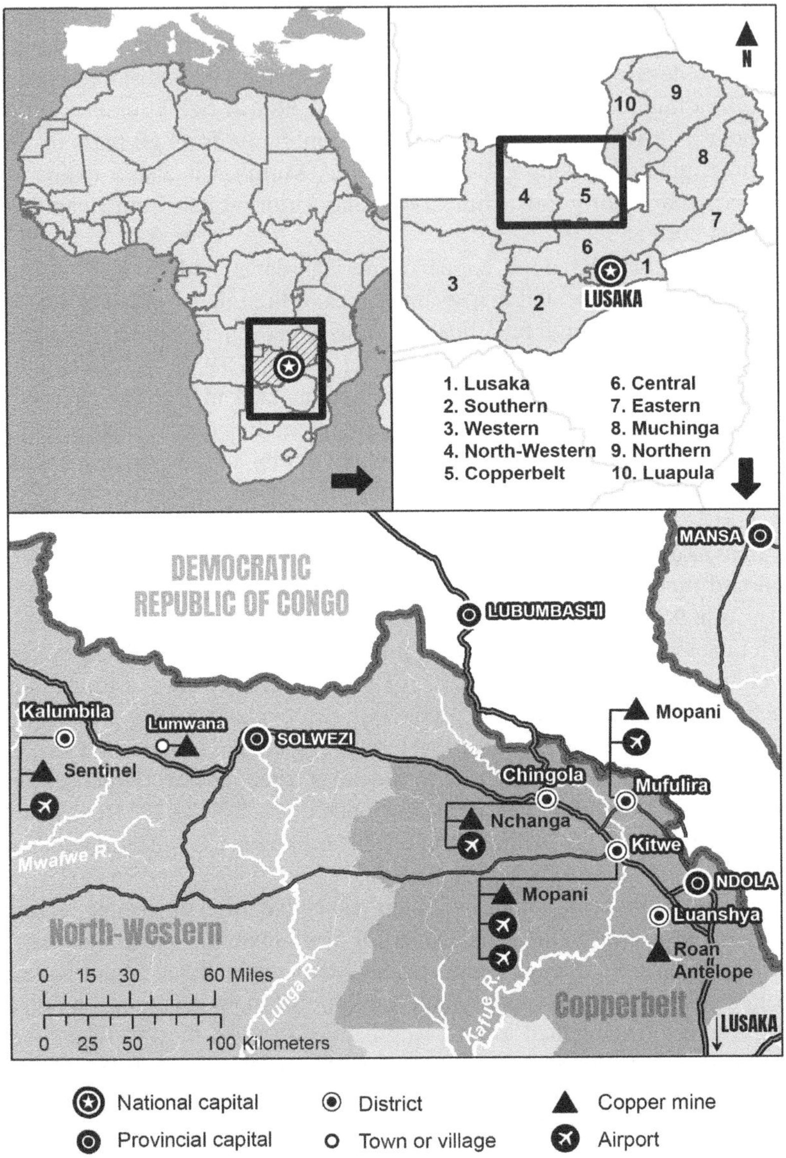

Figure 6.2 Case study sites in North Western and Copperbelt provinces, Zambia. Map by Marilyn Ishikawa.

(Mwale 2015). The arrival of mining in a previously rural area caused a range of economic, social, infrastructural, and environmental changes. To prevent uncontrolled urbanization along Zambia's mining frontier, First Quantum Minerals built an entire town in Kalumbila, including a gated community with a golf resort, new roads, and access to stable electricity. Lumwana mine, a large-scale open pit copper mine run by Barrick Gold, also established a mine township near its operations, albeit more securitized than First Quantum's Kalumbila town. Visitors to the Lumwana mine and estate go through a securitized checkpoint. In-migration, particularly of experienced mineworkers from the Copperbelt, was so rapid that many chose to live in less regulated areas, such as Manyama and Kananga, further away from the mine. Tensions between local residents, traditional authorities, and newly migrated mineworkers have escalated as a result, as access to jobs and profits from mining are heavily contested (Kapesa, Mwitwa, and Chikumbi 2015; O'Callaghan 2019).

By combining analysis of Zambia's established mining towns of Mufulira and Luanshya with a focus on Kalumbila as a newly developed mine town, we can better tease out the features of mobility that emerge in various mining localities.

Mobility in Zambian mining localities: motives and drivers

Throughout the twentieth century, the development of Zambia's copper mines depended on attracting sufficient laborers to work mineral deposits. These labor demands relied on varied forms of mobility. As the Copperbelt was a sparsely populated area, colonial officials and mining companies together set up a system of labor migration, whereby male workers from rural areas several hundreds of kilometers away were enticed to come to the Copperbelt mines on contracts of six to twelve months, before supposedly returning "home" to their families (Larmer et al. 2021). Continuously high labor demands encouraged urban stabilization over the years, but mobility remained high. Particularly after 1945, entire families settled on the Copperbelt and built their lives in its urban centers, forming a strip of settlements with hundreds of thousands of people (Larmer 2021). These mining localities had varied characteristics, in terms of settlement patterns, livelihood activities, and administrative systems. For example, the Copperbelt towns had a dual administration, with some areas administered by mining corporations and others by municipalities. Certain rural peripheries, in contrast, fell under the administration of customary authorities. Our use of "mining localities" reflects such variations in place and place-making processes.

Following several decades of expansion, the Copperbelt mines succumbed to a protracted period of economic crisis starting in the late 1970s. As copper production plummeted, mineworkers lost their jobs and urban residents struggled to "get by" (Mususa 2021). Deborah Potts (2005) has argued that this period of "decline" ushered in a trend of "counter-urbanization," as Copperbelt residents moved to rural areas or to the capital city Lusaka, where economic fortunes appeared brighter. The post-2000 mining boom, which spurred the opening of multiple new mines in Zambia's North Western province, has rekindled debates about the "expectations of modernity" (Ferguson 1999) that accompany mining-induced mobility (O'Callaghan 2019).

Historians and social scientists have conceptualized mining-induced mobility and urbanization in nuanced and insightful ways (see Ferguson 1999; Larmer 2021). Yet, while the assumption that mining is what attracted people to the Copperbelt and other mining hubs is a useful starting point, it does not capture the rich variety of patterns of mobility to and away from mining localities. We argue that it is worthwhile to also consider other dynamics that drive mobility – such as new opportunities for agricultural production, forestry, and trade in mining localities, as well as social ties and networks – to understand how Zambian mining localities have been made. Miles Larmer (2016) has argued, for instance, that in the first half of the twentieth century, the Copperbelt attracted scores of women and men "engaged in economic activities that were both symbiotic with and parasitic on the mine economy – charcoal burning, growing food (illegally) on mine-owned land, beer brewing, and so on." All this economic activity required and enticed forms of mobility, linking the urban Copperbelt to its peri-urban and rural hinterlands, as well as to more distant rural areas. These flows grow and wane within a shifting political economy, as commodity booms and downturns direct people to new livelihood opportunities and networks.

In his 1947 study of Mufulira's market, William Brelsford (1947) mentions produce such as cassava, maize, dried fish, charcoal, and tobacco, some of which were sourced from Congo or from lakes in Luapula, hundreds of kilometers away. While certain traders moved back and forth between distant rural areas and Copperbelt towns, others settled in urban areas but continued to engage in itinerant trade. Mining activity definitely stimulated the emergence of markets, but subsequently agricultural production and engaging in trade could become independent motives for women and men to move to mining localities (Peša 2020). Access to forest products, charcoal in particular, also enticed rural-urban mobility. Urban households depended on charcoal for cooking and heating purposes, and because the rapidly urbanizing Copperbelt caused widespread deforestation of areas adjacent to towns, a vibrant charcoal trade with peri-urban and more distant rural areas soon emerged (Chidumayo 1989). For Dar es Salaam, Emily Brownell (2020, 180) explained that charcoal became "a way to survive the rupturing economy." Particularly during periods of economic crisis, in the 1980s and 1990s, trading charcoal appeared a more feasible livelihood pathway than finding mining employment.

Mobility behavior has also been influenced by social ties, including those revolving around family, marriage, and religion. Rhodes Livingstone Institute anthropologists pioneered the study of social change on the Copperbelt, examining how rural migrants became adjusted to urban life (see Epstein 1958). Rural-urban connections continued to be important throughout the twentieth century, and they did not simply result in "permanent urbanization" (Heisler 1974). Some long-term Copperbelt residents moved to rural regions and the rural peripheries of urban areas after retirement, or after being laid off. Because of the massive redundancies that followed privatization of Zambia's state industries and mines in the 1990s, urban residents straddled both rural and urban lifestyles, increasingly relying on land acquired in rural peripheries for agricultural livelihoods (Smart, Nel, and Binns 2015; Mususa 2021). For many, finding a wife from their "home village" remained important, even if they had never actually resided in rural areas themselves (Peša and Henriet 2021). Religious connections, in addition, linked rural parishes to urban churches and vice versa (Haynes 2017).

Taken together, these observations call into question the characterization of Zambia's urban mining centers as "enclaves" (Ferguson 2005), suggesting instead that mining localities have always functioned as connective "hubs." Mineral extraction, recent scholarship argues, produces "enclave economies" that are "deeply integrated into the global economy" but at the same time "fragmented from national space" (Bridge 2009). In a similar vein, James Ferguson (2005) asserts that capital "hops" between mining localities and multinational investors, skipping over the hinterland. Yet focusing on the spheres of trade, agriculture, and forestry, and on how people developed social ties, shows that Zambian mining localities never were closed-off enclaves. Despite the efforts of mining corporations, both in the past and present, to delineate and close off the areas of their mining operations, electricity networks and road and rail infrastructure developed for the mining industry facilitated wider interconnections, expanding older established rural economies and creating new urban ones (Hönke 2018). Zambia's various mining localities thus functioned as hubs, connecting rural and urban areas, and generating new social, economic, and political ties. The mining "hub" in this regard more closely resembles the analytical space between the industrial agglomeration and the enclave (Phelps, Atienza, and Arias 2015). Here Rita Kesselring's (2018) understanding of the dynamics occurring in Zambia's new extractive locales is useful, describing mining corporations' efforts to create an enclave as a process of trying to close off wider economies and mobilities. As we show through the survey data and life histories presented in the following sections, the mining economies of Zambia's new and older mining regions have never been divorced from other economic activities.

Surveying mobility in a Zambian mining locality: Kananga in Kalumbila district

Prior to the 2000s, Kananga, a settlement in the district of Kalumbila, had been a relatively small village along the Solwezi-Mwinilunga road located opposite the Meheba Refugee Camp (see Figure 6.2). Meheba, a roughly 720-km^2 settlement, had been established in 1971 to host people fleeing the Angolan civil war. In recent years, it has taken in refugees from DR Congo, Burundi, and Somalia (see Neto 2018). Since the early 2000s, with the development of the Lumwana mine, located about 17 km away, significant numbers of people have started to arrive and settle in the area. Like other villages in Zambia located in the proximity of a major road, Kananga had markets along the roadside (which primarily sold agricultural produce from the Meheba refugee camp and its surroundings). New settlements in the area grew outward from the road toward the interior, accessed by a few graded roads; others were created by increasing traffic. While population density remained low overall, by 2021, there were visible signs of prosperity, notably in the form of large new houses built of burnt brick and concrete and roofed with iron sheets, instead of the previous earthen wall buildings that had been common in the area.

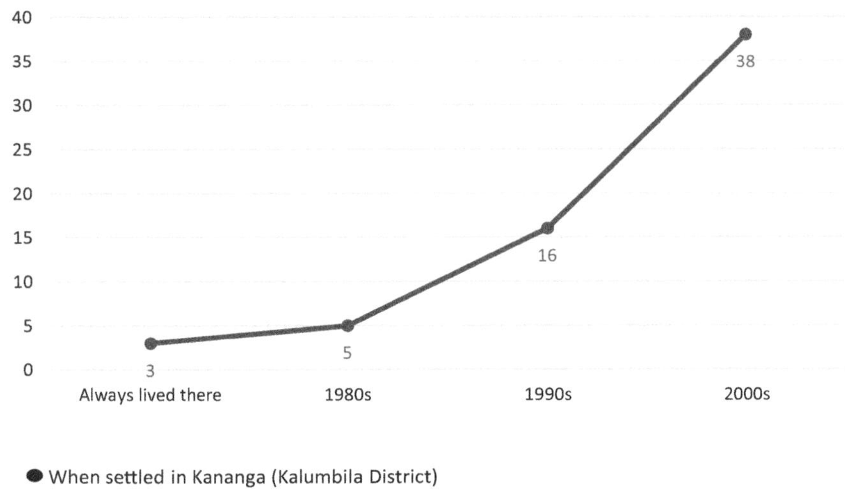

Figure 6.3 When settled in Kananga, Kalumbila district, among survey respondents.

The survey of 62 respondents conducted in October 2021 confirmed the recent arrival of many residents to the settlement of Kananga: 38 respondents had settled in the area after 2000, with the majority (31) arriving after 2006, with the establishment of the Lumwana and Kalumbila mines. Of those surveyed, 21 had arrived in the area between 1988 and 1999 (see Figure 6.3). This was at the height of the nationwide economic crisis that saw job layoffs as industries closed or were privatized. Most layoffs around that time were on the Copperbelt, with redundancies arising from the sale of the state-owned mines. In the town of Mwinilunga in the North Western province, the closure of a fruit-canning factory in 1991 affected not only factory employees but also the agricultural sector that sold the pineapples processed in the cannery. This context highlights how economic fortunes, both good and bad, may influence mobility.

Only three of the respondents surveyed in the settlement of Kananga had been born there (see Figure 6.4). However, the majority had been born within the North Western province (17 in Mwinilunga, 12 in Solwezi, seven in Kabompo, four in Zambezi, three in Mufumbwe), followed by those born in the Copperbelt province (four in Luanshya, three in Kitwe, three in Ndola, two in Chingola, one in Mufulira). The rest had been born in more distant regions (two in Lusaka, one in Kabwe, one in Kalomo, one in Mongu), and one respondent in DR Congo, in Kalemie, near Lake Tanganyika. What this indicates is that regional ties may have played a role in choosing to settle in the area, and potentially for accessing land. After arriving in Kananga, the majority (54 out of 62 respondents) acquired a piece of land, most by buying land (21) or having it granted by customary authorities (21), with others occupying land (five) or inheriting it through kinship (five), and only one acquiring land through the municipal authority (see Figure 6.5). These results indicate an interest in settling in the area long-term, as was confirmed in informal interviews.

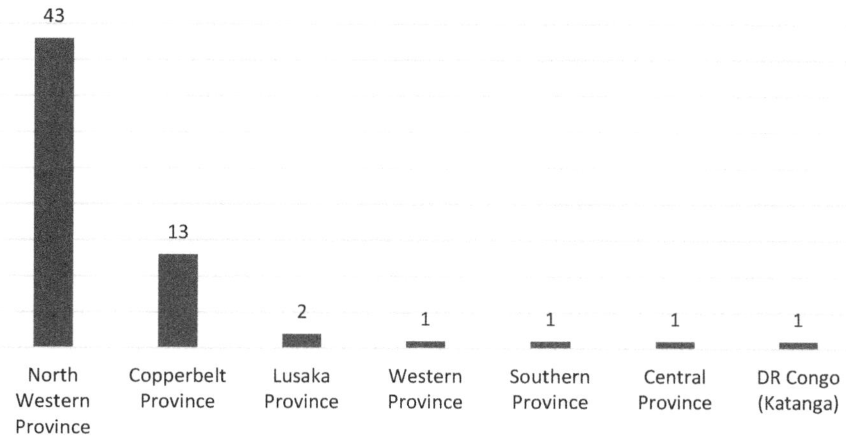

Figure 6.4 Where born, among survey respondents in Kananga, Kalumbila district.

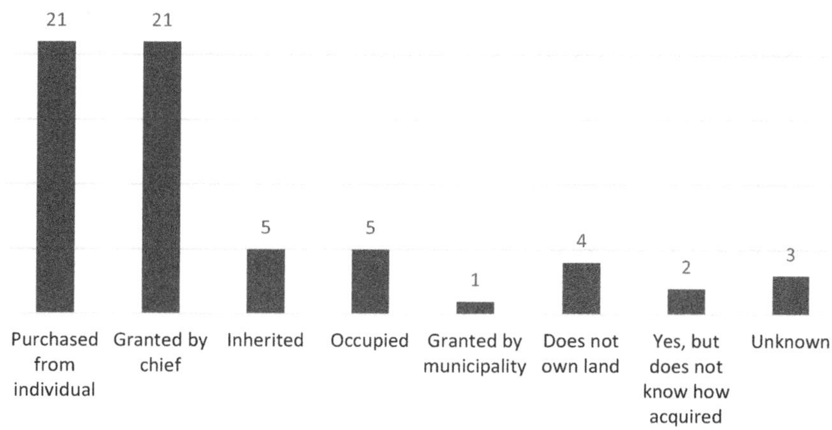

Figure 6.5 Land ownership, and how acquired, among survey respondents in Kananga, Kalumbila district.

Settling in Kananga did not mean, however, that residents' ties to other areas diminished. A total of 52 of those surveyed traveled from the area at least once a year, with 39 of them traveling monthly, primarily for family and business purposes. Most of the respondents' mobility was within the province – to towns such as Solwezi, Mwinilunga, Zambezi, Mufumbwe, Kasempa, and Kabompo – and thereafter to Copperbelt towns, including Chingola, Chililabombwe, Kitwe, Kalulushi, Luanshya, Mufulira, and Ndola. Some also traveled further afield to the capital city Lusaka, the border town of Nakonde, and the towns of Kipushi' in DR Congo, and Luena in Angola.

While the large-scale mines had catalyzed mobility into the area, only 10 of those surveyed worked in mining. The majority (39) worked in the agricultural economy that, according to several interviewees, was booming because of the heightened demand for food from the newcomers and the mines. There was also a burgeoning service economy in the area that included the running of mobile money booths, a crucial service given that 41 of the 62 survey respondents were sending and receiving financial remittances.[4]

One of the key developments mentioned in interviews with residents of Kananga was the building of a clinic in the area, a positive development associated with the presence of the mines. However, unlike the settlement of Manyama, located just outside the Lumwana mine, where sections had electricity, in Kananga, despite relative signs of prosperity, residents were not connected to the electricity grid. Most residents used solar power for lighting and charcoal for cooking. This was an issue of contention, marking the inequalities in the spatial geography of Kalumbila. Kananga was thus a locality, though still dominated by an agricultural economy, that was being profoundly shaped by mining.

By surveying a settlement like Kananga, which is not a mine township per se, nor an area planned by the municipalities, but a village that has grown over two decades, we gain insight into the dynamics of settlement and mobility in areas near new mines. It also offers insight into the range of economic activities occurring and emerging around mining. This directs us to think of Kalumbila district as comprising varying localities that large-scale mining is reshaping. In this regard, the dynamics of mobility in Kalumbila district resonate with those described by scholars writing on the trajectories of mobility in the Copperbelt towns (Larmer 2021). By looking at women's lives and adopting a focus on the household, these studies showed a wider sphere of economic activity and, in turn, mobilities that went beyond men's labor in the mines (Parpart 1986; Evans 2015; Larmer 2021).

In zooming in on the mobility patterns in Kananga, we can present a fragment of the broader population trends occurring in Kalumbila district, which had the highest average annual population growth rate in Zambia between 2010 and 2022. For planners concerned about mining-driven population growth, in areas outside the mine township, the data presented offers some insight into the livelihood activities and the investments residents are making, such as acquiring land. This and the life stories of mobility and mining in the next section point to the importance of directing greater attention to the places and activities that do not directly involve mining, as well as of treating mining localities as hubs that facilitate other kinds of mobilities and livelihood activities.

Life histories of mobility and mining

To illustrate the widely varied trajectories of mobility that brought individuals to mining localities, this section presents selected life histories from Mufulira, Luanshya, and Kalumbila. These illustrate that mining was but one among many factors that attracted people to these mining localities and kept them there.

Mufulira

Leah Mwale was born in the Copperbelt town of Kitwe in 1950, to parents who had moved from Malawi in search of mining employment. Economic diversification was always part of household livelihood strategies, as Leah's mother sold doughnuts at the local market. Leah attended primary school in nearby Kalulushi. After her first marriage to a businessman ended in divorce, she went to Mufulira to visit her brother in 1975. There she met her new husband, who worked for the mines. They married and started living in mine housing, a two-bedroom house with a kitchen and an indoor bathroom. In 1983, her husband had a work accident, and he was discharged from the mines. Due to his work experience, he easily found new employment in an explosives factory in Mufulira. With her first husband, Leah had worked in a restaurant, and they ran a grocery together. When she moved to Mufulira, Leah continued selling merchandise. She would buy clothes from South Africa, Zimbabwe, or Malawi, which she resold on the Copperbelt. Traveling to these distant destinations required planning, social connections, and knowledge of different types of merchandise. While she would buy particular plastic shoes in Malawi, she would focus on scarves and caps in South Africa, and *chitenge* cloth in Zimbabwe. Demand for certain items could switch quite suddenly, so this required flexibility. Leah had personal relationships with suppliers, whom she called in advance to announce her arrival, and she had friends in whose houses she could sleep during business trips. She also used to buy fish from Luapula province to resell on the Copperbelt. She would make the trading trips together with other women, but nonetheless she remembers them as challenging. During one trip to Zimbabwe, she was harassed, and she lost all her merchandise. She was robbed in South Africa, but the strong networks she had with other women traders helped her get back on her feet again. Leah was also active in women's clubs, run by churches and community members. Church activities took her all around the region: she attended meetings in rural Zambia, and in Botswana and Zimbabwe. While Leah's family history involved mining employment, her mobility was informed by trade, religion, and social networks in important ways. Leah Mwale has built strong and varied connections in Mufulira – these allowed her to travel, but she always remained firmly grounded in the Copperbelt.

Mischek Bwalya was born in Zambia's Luapula province in 1947. His father was a bricklayer, educated by missionaries. When his father got a job on the Copperbelt, the family moved to Kitwe and thereafter to Chingola, where Mischek went to school. Upon completing Form V, Mischek was recruited by the mines to work as a clerk. He worked for the mines for several years, but then he moved to a radio station, various private companies, and finally into civil service. Due to salary differences, he moved from job to job and between Kitwe, Ndola, and Mufulira. Mischek worked for the Ministry of Education, commuting from Chingola to Kitwe and Ndola, but following his transfer to Mufulira in 1972, he moved there permanently. Mischek liked Mufulira because the city attracted many fish traders from Luapula province, his native area. He met his wife in Mufulira. She was born in Mansa, but she came to the Copperbelt to be trained as a typist. His wife worked as a trader, a job that took her to Botswana and South Africa to order merchandise such as clothes, beans, and other goods.

Mischek's mobility was not confined to his professional life. At school, Mischek got involved in UNIP (United National Independence Party) politics, and he became a UNIP councilor in Mufulira in 1981. He actively campaigned in his own ward, building a follower base who voted for him. UNIP affiliation allowed Mischek to travel the country, from Lusaka and Livingstone to Samfya and Kasempa. Mischek worked as a councilor until 1995, when he retired to the position of alderman. Mischek continued to cultivate his support base in Mufulira, for example by joining clubs such as the Country club and the football club. In 1986, Mischek started a farm on the Copperbelt, where he grows 25 hectares of maize. This grounds him in Mufulira and ensures his livelihood even after retirement. Relatives from Luapula regularly come to visit him in Mufulira, and he occasionally takes his children to his native village. Mischek has internalized the slogan "One Zambia, One Nation," arguing that regional differences and ethnicity do not matter in Zambia. In our interview, he portrayed Mufulira as a melting pot, where different people come together and where anybody is welcome. Mischek also clearly distinguished between council-run neighborhoods and mine-run neighborhoods. Whereas the mine-run neighborhoods had been better maintained before mine privatization in the late 1990s, this situation reversed thereafter. Council-run areas such as Kamuchanga quickly grew in population after 2000. Kamuchanga also used to have many beerhalls and party places, earning it the nickname "Sugarland," for all the beautiful girls in the neighborhood. Mischek Bwalya's life history reveals how one could build a close attachment to a mining locality, without directly engaging in mining employment oneself. Mischek's engagements in trade, agriculture, and politics cultivated varied social connections that rooted Mischek in Mufulira.

With a vibrant history centered around mining, Mufulira in recent years has struggled to recover economically. In the context of tax avoidance (Sikka and Willmott 2013), environmental pollution allegations, and labor casualization (Niederberger, Kobi, and Haller 2019), the town's mines and smelter, which were majority owned by Glencore, were renationalized in 2020. The resale of the mines back to the state, though popular and pushed by the mining union, was broadly seen to have occurred under terms that advantaged Glencore and saddled the Zambian state with a large debt (Musonda 2023). This has frozen mining-related activity in the town, giving prominence to other sectors, and has politicized residents' views of mining.

Luanshya

Brian Chisokota worked and lived primarily in Zambia's capital city Lusaka but spent a couple of weekends a month in Luanshya, where his wife, their two primary school-aged children, and a teenage nephew lived. He considered Luanshya his main residence even though he had a permanent job as an accountant with an auditing firm in Lusaka. The older of two siblings, Brian was born in 1973 in the Copperbelt town of Chingola, where his father worked as a miner at Nchanga copper mines, and his mother in seasonal small-scale farming as well as in the occasional trade of agricultural produce like beans and groundnuts. When his father retired from the mines in 1985, his parents moved to Mbala in Northern Zambia, where they acquired a farm on customary tenure and made a living through subsistence farming. Brian and his younger brother stayed with relatives in Chingola to complete their education.

In 1996, after graduating from the University of Zambia in Lusaka where he studied engineering and industrial science, and marrying shortly thereafter, Brian got a job with the mines in Luanshya. He was offered accommodation in the low-density mine township. However, a few years into his job, he was laid off in the redundancies that ensued following the privatization of the Luanshya mines that had been part of the state mining conglomerate, Zambia Consolidated Copper Mines. As with other retrenched workers, the mines were unable to pay out Brian's redundancy package, and thus they offered him the opportunity to buy a mine house in lieu of the full payment. Rather than buy the apartment he was staying in at subsidized rates, he opted to buy a large, three-bedroom house not far from the town's golfing greens, which was at the time being rented by a teacher. Brian reported that "the ministry of education never came through for the teacher," who had hoped to buy it as a sitting tenant.

Following his retrenchment from the mines, Brian retrained as an accountant, and in 2000, he got a job in Ndola with an accounting firm, commuting daily from Luanshya. A few years later, in 2004, he got a higher paying job with an accountancy firm in Lusaka, which is about 320 km from Luanshya. However, like some other residents, he opted not to move his family to Lusaka, highlighting that the cost of living in the capital city was higher, and the quality of life and services such as education would be better for his family in Luanshya on the income he was earning. His two children were attending primary school at the former mine trust schools, which were vaunted for their high-quality early education. Brian on average traveled twice a month from Lusaka to visit his family, and once a year to visit his parents in Mbala. Asked if he would consider retiring to Mbala, or the countryside, as his parents did, he said no, as rural life was something that he did not see himself becoming accustomed to. However, Brian and his wife Tricia, who has an administrative job in Kitwe and commutes daily from Luanshya, had acquired a farm in the Kafubu farming block in the rural peripheries of Luanshya, where they keep a small herd of cattle and grow maize. The farm is mainly managed by Brian's in-laws, who live in the house they built at the farm. Brian and his wife also invested in a second property in the low-density section of Luanshya that was historically managed by the municipality, as the town, like other Copperbelt towns had run a dual administration: one of the mine, the other of the municipal area. Luanshya, for the Chisokota family, can be seen as a base from which they run their household and its economies, keeping livelihood options open, be these in farming or in real estate, and being able to be mobile to pursue stable formal employment, should it be available.

Luanshya was at one time in the 2000s considered to be the most economically depressed town on the Copperbelt, with predictions that it would become a ghost town as its population growth slowed. It has since had a modest rebound on the back of a growing agricultural economy and renewed investment in an open-pit mine by the Chinese firm, China Non-Ferrous Metals Corp. Though dominated by mining in its heyday, it had – and, though at a reduced scale, still has – a manufacturing sector, including copper wire production, steel fabrication, and stockfeed production, that is expected to grow due to its location in proximity to key transport hubs. The sale of housing to sitting tenants on the Copperbelt following the privatization of the mines allowed residents such as the Chisokotas to seek opportunity in other areas during the crisis and at the same time make investments in Luanshya itself.

Kalumbila

After 2010, the municipal authority in Kalumbila started to generate plans for longer-term settlement in the area. Among the newcomers, some moved to Kalumbila for job opportunities with the mines. Many, however, found other livelihood options or they used earnings from mining – after having completed several temporary contracts – to engage in other economic activities.

Michael Mulenga, who was born in 1990 in the Copperbelt town of Chingola, and who worked as a mine safety supervisor at Lumwana mine, was one of them. While he was initially accommodated in Lumwana's heavily securitized mine residential area, where he shared a two-bedroom prefabricated house with a colleague, he soon acquired a residential plot outside of the newly established Kalumbila municipal authority. There he planned to build a house. He had also bought a five-acre piece of land in the agricultural zone near the Kananga settlement, where he employed occasional workers to grow crops.

The main reasons he gave for acquiring land for housing and farming outside of the mining area included considerations of earning a living once his contract ended, as well as the restrictions on mobility[5] that living in the Lumwana mine residential area engendered. He was not alone in this. As a result of the level of control the mine exerted on workers and their visitors' mobility into the mine area, most preferred to live in the adjacent growing settlement of Manyama, which straddled the region's main road outward into the countryside. As a result of the presence of the mine, Manyama had electricity, and consequently, some of the district's most thriving business activities were located there. One could find mobile banking booths, carpentry businesses making door frames and household furniture, hairdressers, restaurants, bars, and guesthouses, as well as schools in Manyama. It was in this area that the municipality established its offices. However, Michael preferred to stay in the mine township when he was on duty, visiting his friends occasionally at the end of the month and his parents, who lived in Kitwe, when he had time off. Though not as mobile as those staying outside the mining settlement, he was nonetheless establishing a livelihood safety net beyond mine employment, through farming. He grew beans and maize on his five-acre plot but had ambitions to establish a small dairy farm, if he managed to acquire more land. Michael's plan for the residential plot he had bought was to rent out a few rooms, something that several residents were already doing in response to the high demand for accommodation in the area.

Mining activity has created a boom in the wider Kalumbila economy, including in Kananga (where the survey was conducted) toward the southeast, located next to the Meheba refugee camp. The long-established refugee camp has become a node for settlement, as its schools and healthcare facilities can be accessed by those who live in the area outside it. It also attracts state aid and development projects that have spurred an agricultural economy, which has contributed significantly to the region's agricultural output. This agricultural production now also feeds the mine's workers. Residents say farming in the area has scaled up. Farmers are reorganizing into cooperatives to pool inputs such as fertilizer subsidies and to gain better market access to the mines and to larger-scale retailers establishing themselves in the area.

Mining has also lent a visibly cosmopolitan character to the area, with mineworkers and other residents from farther afield, such as Australia and South Africa. This is not to say that the region itself had not long been cosmopolitan and connected to wider geographies than those within Zambia. The Meheba refugee camp has seen the long-term residence of people from Angola, the DR Congo, Burundi, and Somalia; many of whom have international diasporic ties.

The region's long-term residents and its newcomers have thus been mobile, within the region and beyond, moving for business and family visits, to access services in other areas and to explore general opportunities. The employment contracts with the mines and their various contractors have tended to be relatively short-term, if one compares them to those of the formerly state-run mines located on the Copperbelt. Immediately after independence, workers had jobs for life, being mobile only across the companies' various divisions located in different Copperbelt towns, though some did change jobs and move to other sectors, or did so after retirement. In the context of the more precarious employment conditions that characterize Zambia in the twenty-first century, the residents of the new and older mining regions no longer see mining as a stable and stabilizing employer. Rather, they see it, as in Michael Mulenga's case, as a sector that provides a period of employment with reasonable earnings that can subsequently be invested in other areas, such as agriculture, real estate, trade, finance, or transportation services. That is why although mining does attract new settlements, spurring investments in land and property, residents remain mobile, maintaining economic and family ties well beyond the mining region. This pattern of work-life mobility is remarkably similar to that in the Copperbelt towns of Mufulira and Luanshya that were established almost a century earlier than Kalumbila.

Conclusion

The making of mining localities in Zambia is shaped not just by mining but also by agriculture, trade, and social activities. These have tended to be made invisible in the narratives of mineral extraction in the country, which portray mining as somewhat disconnected from other sectors. In this regard, large-scale mining operations in Zambia, though displaying some of the characteristics of an enclave, function more as hubs, bringing together a conglomeration of multiple livelihood activities, lifestyles, and mobilities. These, in turn, co-produce varying localities, and the varied character of places – from the planned and regulated mine township to trade and agricultural settlements in peri-urban and rural areas. The making of mining localities thus goes beyond the mine township planned for the settlement of workers.

Throughout this chapter, we have shown how residents maintain their interconnections with other places for family, economic, and social reasons, thus generating mobilities within and beyond the places where they live. In the life histories collected from both the new mining district of Kalumbila and the older mining towns of Mufulira and Luanshya, residents are investing in places and acquiring land and assets, not only in the places they live but also where they have social and economic ties. They are involved in trade that sees them travel within the country and abroad. They are also maintaining social ties, not only by visiting family members but also through remittances of cash and food. The survey conducted in Kananga confirmed these mobilities.

In Kalumbila, in the North Western province, where large-scale industrial mining attracts much larger amounts of capital compared to the subsistence agriculture sector that employs much of the people, we posit that mining has turbocharged mobilities and processes of settlement consolidation. The investments in electrification, roadways, railways, and other nodal infrastructure, such as telecommunications, while developed primarily for the mining sector, facilitate wider interconnections, expanding both older established economies and creating new ones (Schubert, Engel, and Macamo 2018). In places such as Luanshya and Mufulira in the Copperbelt province, old and renewed investments in physical infrastructures, including housing, have meant that even in periods of economic downturn, the towns' residents have been able to leverage these to transform livelihood activities (Mususa 2021).

In drawing from cases from the new mine settlements of Kalumbila and the older established towns of Mufulira and Luanshya, we have sought to demonstrate the durability of mobility in mining localities beyond the periods in which they are developed. What shifts, the cases highlight, are the itineraries of their residents. As changes in the political economy occur, these affect what people can do, and how and where they move. Taking an expansive view of mobility, one that pays attention to various indeterminate movements and their varied motivations, allows us to better grasp how mining localities are made through mobility.

Notes

1 The survey administrators (all with high school certificates) were fluent in English and three languages widely spoken in the area, namely Lunda, Kaonde, and Bemba.
2 The interviews were conducted outdoors, keeping social distance. Survey administrators wore face masks and made face masks available to respondents.
3 This chapter draws on research supported by the Riksbankens Jubileumsfond, grant number P19-0271:1, "Changing Urban Residency: Migration, Temporary Settlement and New Urbanisms in Africa."
4 Ten respondents reported working in the service sector, ten as civil servants, and the rest combined agriculture with trade.
5 All residents and visitors must go through a security checkpoint every time they enter or leave the mine area. Their identity documents are checked, bags are searched, and biometric data are collected.

References

Auty, Richard. 2006. "Mining Enclave to Economic Catalyst: Large Mineral Projects in Developing Countries." *The Brown Journal of World Affairs* 13 (1): 135–145.

Brelsford, William Vernon. 1947. *Copperbelt Markets: A Social and Economic Study.* Lusaka: Government Printer.

Bridge, Gavin. 2009. "The Hole World: Spaces and Scales of Extraction." *New Geographies* 2: 43–48.

Brownell, Emily. 2020. *Gone to Ground: A History of Environment and Infrastructure in Dar es Salaam.* Pittsburgh: University of Pittsburgh Press.

Bryceson, Deborah, and Danny MacKinnon. 2012. "Eureka and Beyond: Mining's Impact on African Urbanisation." *Journal of Contemporary African Studies* 30 (4): 513–537.

Bryceson, Deborah Fahy, Katherine V. Gough, Jesper Bosse Jønsson, Crispin Kinabo, Michael Clarke Shand, Cristina Udelsmann Rodrigues, and Paul W.K. Yankson. 2022. "Mineralized Urbanization in Africa in the Twenty-first Century: Becoming Urban Through Mining Extraction." *International Journal of Urban and Regional Research* 46 (3): 342–369.

Castillo, Gerardo, and David Brereton. 2018. "Large-Scale Mining, Spatial Mobility, Place-Making and Development in the Peruvian Andes." *Sustainable Development* 26 (5): 461–470.

Central Statistical Office (CSO). 2012. *Population and Housing Census 2010: Analytical Report.* Lusaka: Central Statistical Office.

Chidumayo, Emmanuel N. 1989. "Land Use, Deforestation and Reforestation in the Zambian Copperbelt." *Land Degradation & Development* 1 (3): 209–216.

Epstein, Arnold Leonard. 1958. *Politics in an Urban African Community.* Manchester: Manchester University Press.

Evans, Alice. 2015. "History Lessons for Gender Equality from the Zambian Copperbelt, 1900–1990." *Gender, Place & Culture* 22 (3): 344–362.

Ferguson, James. 1990. "Mobile Workers, Modernist Narratives: A Critique of the Historiography of Transition on the Zambian Copperbelt." *Journal of Southern African Studies* 16 (3): 385–412.

Ferguson, James. 1999. *Expectations of Modernity: Myths and Meanings of Urban Life on the Zambian Copperbelt.* Berkeley: University of California Press.

Ferguson, James. 2005. "Seeing Like an Oil Company: Space, Security, and Global Capital in Neoliberal Africa." *American Anthropologist* 107 (3): 377–382.

Gluckman, Max. 1960. "Tribalism in Modern British Central Africa." *Cahiers d'Études Africaines* 1: 55–70.

Haynes, Naomi. 2017. *Moving by the Spirit: Pentecostal Social Life on the Zambian Copperbelt.* Berkeley: University of California Press.

Heisler, Helmuth. 1974. *Urbanization and the Government of Migration: The Inter-Relation of Urban and Rural Life in Zambia.* London: Hurst.

Hönke, Jana. 2018. "Beyond the Gatekeeper State: African Infrastructure Hubs as Sites of Experimentation." *Third World Thematics* 3 (3): 347–363.

Kapesa, Robby, Jacob Mwitwa, and Donald Chikumbi. 2015. "Social Conflict in the Context of the Development of New mining Concessions in Zambia." *Southern African Peace and Security Studies* 4 (2): 41–62.

Kesselring, Rita. 2018. "At an Extractive Pace: Conflicting Temporalities in a Resettlement Process in Solwezi, Zambia." *The Extractive Industries and Society* 5 (2): 237–244.

Larmer, Miles. 2016. "At the Crossroads: Mining and Political Change on the Katangese-Zambian Copperbelt." *Oxford Handbooks Online.* doi: 10.1093/oxfordhb/9780199935369.013.20.

Larmer, Miles. 2021. *Living for the City: Social Change and Knowledge Production in the Central African Copperbelt.* Cambridge: Cambridge University Press.

Larmer, Miles, Enid Guene, Benoît Henriet, Iva Peša, and Rachel Taylor, eds. 2021. *Across the Copperbelt: Urban & Social Change in Central Africa's Borderland Communities.* Oxford: James Currey.

Luning, Sabine. 2018. "Mining Temporalities: Future Perspectives." *The Extractive Industries and Society* 5 (2): 281–286.

MacMillan, Hugh. 1993. "The Historiography of Transition on the Zambian Copperbelt - Another View." *Journal of Southern African Studies* 19 (4): 681–712.

Massey, Doreen. 1993. "Questions of Locality." *Geography* 78 (2): 142–149.

Mickels-Kokwe, Gun. 2006. *Small-Scale Woodland-Based Enterprises with Outstanding Economic Potential: The Case of Honey in Zambia.* Bogor, Indonesia: Centre for International Forestry Research.

Mitchell, J. Clyde. 1966. "Theoretical Orientations in African Urban Studies." In *The Social Anthropology of Complex Societies,* edited by Michael Banton. London: Tavistock Publications.

Musonda, James. 2023. "He Who Laughs Last Laughs the Loudest: The 2021 *Donchi-Kubeba* (Don't Tell) Elections in Zambia." *Review of African Political Economy.* doi: 10.1080/03056244.2023.2190452.

Mususa, Patience. 2021. *There Used to Be Order: Life on the Copperbelt after the Privatisation of the Zambia Consolidated Copper Mines.* Ann Arbor: University of Michigan Press.

Mutamba, Manyewu. 2007. "Farming or Foraging? Rural livelihoods in Mufulira and Kabompo districts of Zambia." Center for International Forestry Research and Rhodes University, 20.

Mwale, Chimwemwe. 2015. "Solwezi Gets 2 New Districts." *Zambia Daily Mail,* August 29.

Negi, Rohit. 2014. "'Solwezi *Mabanga*': Ambivalent Developments on Zambia's New Mining Frontier." *Journal of Southern African Studies* 40 (5): 999–1013.

Neto, Pedro Figueiredo. 2018. "Ambiguous Boundaries between Exclusion and Inclusion: Experiences from the Meheba Refugee Camp (Zambia)." In *Changing Societies: Legacies and Challenges. Vol. 1. Ambiguous Inclusions: Inside Out, Inside In,* edited by Sofia Aboim, Paulo Granjo, and Alice Ramos, 29–47. Lisbon: Imprensa de Ciências Sociais.

Niederberger, Thomas, Madlen Kobi, and Tobias Haller. 2019. *The Open Cut: Mining, Transnational Corporations and the Commons*. London: Routledge.

Nyamazana, Mushiba, Grayson Koyi, Patricia Funjika, and Edward Chibwili. 2017. *Zambia Refugees Economies: Livelihoods and Challenges*. Geneva: UNHCR.

O'Callaghan, Margaret. 2019. 'Copperfields: A History of the Impact of the First Decade of a Mining Boom in North Western Province Zambia, Circa 2002-2015." Self published. https://margocall.files.wordpress.com/2020/03/copperfields-r2-with-references.pdf.

Parpart, Jane L., 1986. "The Household and the Mine Shaft: Gender and Class Struggles on the Zambian Copperbelt, 1926–64." *Journal of Southern African Studies* 13 (1): 36–56.

Peša, Iva. 2020. "Crops and Copper: Agriculture and Urbanism on the Central African Copperbelt, 1950-2000." *Journal of Southern African Studies* 46 (3): 527–545.

Peša, Iva. 2022. "Copper Communities on the Central African Copperbelt 1950-2000." In *Born with a Copper Spoon: A Global History of Copper, 1830–1980,* edited by Robrecht Declercq, Duncan Money, and Hans Otto Frøland, 527–545. Vancouver: UBC Press.

Peša, Iva, and Benoît Henriet. 2021. "Beyond Paternalism: Pluralising Copperbelt Histories." In *Across the Copperbelt: Urban & Social Change in Central Africa's Borderland Communities,* edited by Miles Larmer, Enid Guene, Benoît Henriet, Iva Peša, and Rachel Taylor, 27–51. Oxford: James Currey.

Phelps, Nicholas A., Miguel Atienza, and Martin Arias. 2015. "Encore for the Enclave: The Changing Nature of the Industry Enclave with Illustrations from the Mining Industry in Chile." *Economic Geography* 91 (2): 119–146.

Potts, Deborah. 2005. "Counter-Urbanisation on the Zambian Copperbelt? Interpretations and Implications." *Urban Studies* 42 (4): 583–609.

Rubbers, Benjamin. 2019. "Mining Towns, Enclaves and Spaces: A Genealogy of Worker Camps in the Congolese Copperbelt." *Geoforum* 98: 88–96.

Schubert, Jon, Ulf Engel, and Elísio Macamo, eds. 2018. *Extractive Industries and Changing State Dynamics in Africa: Beyond the Resource Curse.* London: Routledge.

Schumaker, Lyn. 2001. *Africanizing Anthropology: Fieldwork, Networks, and the Making of Cultural Knowledge in Central Africa.* Durham, NC: Duke University Press.

Schumaker, Lyn. 2008. "Slimes and Death-Dealing Dambos: Water, Industry and the Garden City on Zambia's Copperbelt." *Journal of Southern African Studies* 34 (4): 823–840.

Sikka, Prem, and Hugh C. Willmott. 2013. "The Tax Avoidance Industry: Accountancy Firms on the Make." *Critical Perspectives on International Business.* Centre for Global Accountability, University of Essex.

Straube, Christian. 2021. *After Corporate Paternalism: Material Renovation and Social Change in Times of Ruination.* New York: Berghahn.

Smart, Jessie. 2015. "Urban Agriculture and Economic Change in the Zambia Copperbelt: The Cases of Ndola, Kitwe and Luanshya." PhD diss., University of Otago.

Smart, Jesse, Etienne Nel, and Tony Binns. 2015. "Economic Crisis and Food Security in Africa: Exploring the Significance of Urban Agriculture in Zambia's Copperbelt Province." *Geoforum* 65: 37–45.

Udelsmann Rodrigues, Cristina, and Ana Paula Tavares. 2012. "Angola's Planned and Unplanned Urban Growth: Diamond Mining Towns in the Lunda Provinces." *Journal of Contemporary African Studies* 30 (4): 687–703.

Udelsmann Rodrigues, Cristina, Patience Mususa, Karen Büscher, and Jeroen Cuvelier. 2021. "Boomtown Urbanization and Rural-Urban Transformation in Mining and Conflict Regions in Angola, the DRC and Zambia." *Sustainability* 13 (4): 2285.

Umar, Bridget. 2017. "Conservation Agriculture Promotion and Uptake in Mufulira, Zambia-A Political Agronomy Approach." *Journal of Sustainable Development* 1 (1): 156–169.

Wilson, Godfrey. 1941. *An Essay on the Economics of Detribalization in Northern Rhodesia.* Manchester: Manchester University Press.

Zambia Statistics Agency (ZSA). 2022. *Census of Population and Housing -2022: Preliminary Report*. Lusaka: Zambia Statistics Agency.

7 The governance of ASGM in Guinea and Côte d'Ivoire

(Im)mobility, territory, and technological change

Anna Dessertine, Robin Petit-Roulet, Muriel Champy and Ibrahima Kalil Doumbouya

Introduction

In recent decades, the worldwide surge in artisanal and small-scale gold mining (ASGM) has been particularly visible in West Africa, where several million people now live directly or indirectly from this activity (OECD 2018). In this area, ASGM is an ancient activity, which contributed to the power and influence of medieval African empires. But the spectacular rise in world gold prices since the 2000s has caused an intensification of the activity in these historic areas, leading to the discovery and exploitation of new deposits and to a surge of local, national, and international mobility directed toward gold mining areas. More recently, mechanized forms of extraction and processing of ore have been developed, to the point that distinguishing between AGSM and (semi-)industrial mining is more a question of the legal status of a site than of techniques of extraction (e.g., artisanal vs. mechanized). Indeed, formal industrial and semi-industrial mines are multiplying in parallel with the development of mechanized and non-mechanized artisanal and informal gold mining.

In contrast to active sites of industrial mining, access to gold deposits in ASGM sites in Guinea and Côte d'Ivoire (and also Mali, Senegal, and Burkina Faso) is open to anyone, regardless of origin, nationality, or ethnicity. Workers live under the jurisdiction of customary authorities, whose rules appear relatively unrestrictive and similar across the region. Building on a set of norms, values, and work processes shared by gold miners in West Africa, the "social ethic of the mine" (Panella 2007) also facilitates mobility between the different sites in the sub-region (Grätz 2003; Bolay 2017).

Even though the historical integration of gold mining into the rural economy is increasingly challenged by the socio-environmental repercussions of ASGM (Ouédraogo 2019; Ngom et al. 2022), many families combine ASGM and agriculture, transferring labor and capital from one activity to the other (Hilson and Garforth 2012; Pijpers 2014; Hilson 2016; Lahiri-Dutt 2018; Brugger and Zanetti 2020; Huntington and Marple-Cantrell 2022). This traditional coexistence has led some analyses to adopt a binary framing in which "mobile" ASGM is opposed to "immobile" agricultural activities, with research on ASGM putting a strong focus

DOI: 10.4324/9781003313236-9

on mobility (Grätz 2003, 2004; Dessertine 2013; Bolay 2016, 2022; Grégoire and Gagnol 2017; Chevrillon-Guibert, Gagnol, and Magrin 2019). In this chapter, we argue that instead of adopting this binary framing to understand patterns of labor migration in areas of ASGM, these mobilities should be analyzed in relation to spatially fixed systems of territorial governance. How are contemporary ASGM mobilities affecting these systems of power relations? And how are they affected by them?

Two conceptual propositions inform our analysis. The first is that the understanding of mobility must take immobility seriously into account (Cresswell 2010), because mobilities need "moorings" (Hannam, Sheller, and Urry 2006). The study of ASGM mobilities cannot be extracted from the territories these mobilities link together, since mobilities and territories are in a dialectic relationship. Indeed, the role of mobility in the constitution of territories and in the emergence of mining sites – which can be considered the results of "interruptions" in ASGM mobilities – cannot be ignored. Relatedly, the second proposition is that governance in its territorial dimension is central to the regulation of mobility. Our approach to governance – understood as the modalities of administration of an object – is descriptive rather than normative.

> Governance is conceived as a set of interactions (conflict, negotiation, alliance, compromise, avoidance etc.) resulting in more or less stabilized regulations, producing order and/or disorder (the point is subject to diverging interpretations between stakeholders) and defining a social field, the boundaries and participants of which are not predefined
>
> (Blundo and Le Meur 2009)

In our study areas, the rules organizing mobility link precedence to political power: firstcomers on a given site integrate newcomers by placing them under their political and economic tutelage (Kopytoff 1987). This historically constituted power relationship persists until now. Power structures are therefore shaped by mobilities and benefit from them. We thus analyze mobility and territorial governance as two sides of the same coin (Boyer 2013), considering that the mobility of some is made possible (or at least regulated) by the immobility of others. In our case study, mobility is linked to resource exploitation. Therefore, the resource governance that regulates mining is coupled with a territorial governance and a mobility governance.

In order to analyze the interaction between ASGM mobility and territorial governance, this chapter takes a comparative approach, focusing on five mining sites in three gold mining areas in two neighboring countries with differing mining histories (see Figure 7.1). It is based on ethnographic field data collected by the authors in Guinea (16 months between 2011 and 2019, seven months between 2021 and 2022) and in Côte d'Ivoire (two months in 2022). Interviews were conducted with gold miners and their families as well as with people locally in charge of the regulation of gold mining. Several types of data were collected: life stories, statements on the modalities of exploitation (access to resources, techniques, mobility, income,

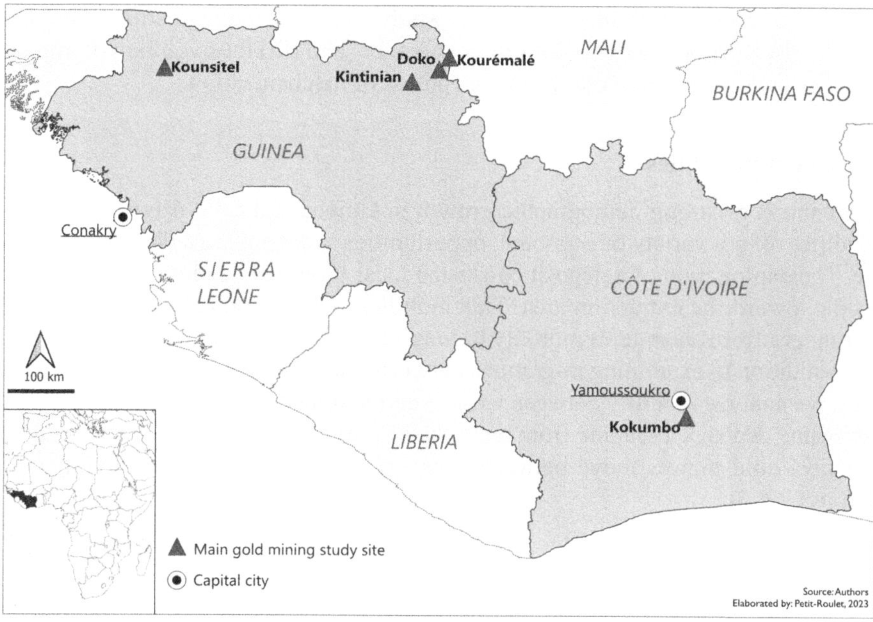

Figure 7.1 Gold mining study sites in Guinea and Côte d'Ivoire. Map by Robin Petit-Roulet.

work organization), and information on the regulation of the activity (governance actors, rules and principles of regulation, power relations between actors).

In northeastern Guinea, ASGM is a centuries-old activity: the recent rise in gold prices has only led to its intensification and to its progressive mechanization. On the contrary, the Gaoual region in north-western Guinea is a new mining area. The deposit was only discovered and exploited in 2021, using both manual and mechanized techniques. In Côte d'Ivoire, social and economic life in the Yamoussoukro area has revolved around the plantation economy since the colonial period, even though some mining sites existed in the nineteenth century. The main study site in Côte d'Ivoire, Kokumbo, had indeed developed around traditional gold panning performed by the Baoule, followed by colonial industrial exploitation of gold deposits at the end of the nineteenth century (Chauveau 1979). But the plantation economy favored by the Houphouët-Boigny regime led to the slowing down, and even the cessation, of the activity. When gold mining re-emerged in the 2010s in a globalized context of a gold boom, it was immediately mechanized.

Although in different forms, including because the logic of individual land ownership is more deeply rooted in Côte d'Ivoire than in Guinea, we observe that mobilities in both countries paradoxically reinforce the rhetoric of autochthony and the political power associated with spatial fixity. In other words, the power of the "immobile" power structures grows through the "mobile" workers. This reinforces the distinction between natives and foreigners, and the importance of being recognized as a native. To explore this idea, this chapter will describe successively (i) the diversity of mining mobilities in our case study areas, (ii) the regulation of mining mobility by territorialized systems of governance, and (iii) the evolution of mining mobility and mining governance in the context of mechanization.

A world of mobilities

In a context of strong demographic growth in Guinea and Côte d'Ivoire,[1] mining localities offer a variety of economic opportunities accessible to a diversity of people. The exploitation of a deposit boosts the local economy, leading to an influx of people toward the extraction area. This mobility is distributed along a continuum that ranges from short-term mobility to long-term settlement in mining areas.

In addition to examining migrations between non-mining areas and gold mining sites, we analyze mobility between various extraction sites. Searching for the most promising deposits, escaping from repression by armed forces, or following rumors of gains, gold miners move between sites, connecting them into an archipelago (Arrault 2005).

The multiple temporalities of mining mobility: from seasonal migrations to settlement

Mining migrations in Guinea and Côte d'Ivoire differ in terms of duration and rhythm. In Guinea, ASGM attracts people from across the country. Most of them move to ASGM mining areas for only a limited period (usually over the dry season). Some other miners are inhabitants of the few villages that are located in the richest gold areas. They exploit gold directly on their lands, such as in the Bouré region. Interviews conducted in some of these villages have shown that artisanal gold mining has been preferred to agriculture for at least six generations (long before the gold boom of the 2000s).

Although they regularly travel to mining sites for a few weeks or a few months each year, most miners still consider their village of origin to be their main residence. The limited time they spend in the artisanal mines also prevents them from accessing specialized activities. Men are usually involved in vertical shaft digging, while women help to extract bags of ore.

People with different profiles practice this short-term artisanal mining mobility. For some gold miners, these migrations are a way to pay off a debt or finance an investment. In the case of young men, these movements are similar to other juvenile labor mobilities that are aimed at enabling the accumulation of financial capital, while contributing to the individual's self-realization and access to adult status (Dessertine 2013; Bolay 2017). In Guinea, this mobility has fueled different economic sectors throughout history: mobility toward diamond mining regions was predominant between the 1970s and the 2000s but is now directed toward gold mining sites.

This short-term mining mobility is also practiced in Côte d'Ivoire, but most miners interviewed in the area of investigation are involved in migrations that are longer-term (more than a year) and less opportunistic (as we will see in regard to the supervised mobility discussed below), particularly concerning young men coming from Burkina Faso. In Burkina-Faso, juvenile labor migration is an integral part of the masculine social life cycle (Thorsen 2006; De Lange 2007; Champy 2022) and is part of a historical pattern of migration toward Côte d'Ivoire (Champy 2015), which has its roots in the colonial period, when the Upper Volta (renamed Burkina Faso in 1984) was used as a manpower reservoir for the development of Côte d'Ivoire.

In these two countries, whether through autonomous or supervised forms of mobility (see below for discussion of the latter), gold miners generally try to use their mining income to start other activities after a few months or years. Gold mining is mainly considered a temporary activity, to be undertaken in a period of transition. However, in some cases, the miners do not return to their home village. For this group, the mines are a way to finance other activities, notably settlement in a big city. Other people stay involved in gold mining activities for several years, sometimes decades. The accumulation of experience allows them access to more remunerative and specialized forms of work, such as the exploitation of underground galleries. Gold mining is the main activity in terms of time and income for these people who have become "professional" gold miners. Their mobility no longer entails going back and forth between their place of origin and the mining zones. Instead, they migrate over the long term to the mining area and set up their main residence there. Some also maintain dual residencies, one in the mining locality, and the other in their village of origin (Bolay 2017).

Independent and dependent mining mobilities

Moving to a gold mining area has several advantages. Because it provides access to income opportunities that are often higher than those available in people's areas of origin, becoming involved in ASGM is a way of accumulating capital over a short period in order to finance an investment, cover a debt, pay for studies, etc. In addition, mobility allows people to escape from village and family control and to gain greater autonomy in the management of their income. Most of the people in our three study areas did not originate from the localities where the gold mines were located.

In Côte d'Ivoire, opportunistic migration coexists with a pattern of what we have called supervised mobilities, especially around mechanized sluices, which employ mostly young Burkinabe men, aged 13–20. The same is true for the cyanidation sites, where some workers are as young as ten years old. This work is not considered to be physically challenging – even though prolonged contact with the cyanide salts and solutions is harmful to one's health.

The relative success of generations of Burkinabe who settled in Côte d'Ivoire over the decades indeed relied on their ability to mobilize an available workforce from their home villages, using the promise of the initiatory trip to the neighboring country as an inducement: in many rural areas, you were not considered a man if you hadn't been to Côte d'Ivoire. Labor migration from Burkina Faso to Côte d'Ivoire was historically oriented toward the agricultural sector, primarily plantations. For several decades, some Burkinabe migrants have been settling in Côte d'Ivoire, building up an important network, and establishing workforce circulation channels to support the plantation economy, in the process contributing to the Ivoirian El Dorado's power of attraction. Since the gold boom, these labor circuits have been redirected from the plantations to the mines. Many Burkinabe migrants, being already familiar with gold digging back home, also consider this activity less challenging and more profitable than plantation work.

The gold labor circuits in Kokumbo are generally managed by Burkinabe men who entered ASGM a few years back and were able to accumulate sufficient capital and experience to buy their own engine. Claiming that children born in Côte d'Ivoire are not as hard-working, they prefer recruiting workers from their own extended families or villages of origin (or areas nearby to these). They organize and pay for the transport of these workers, often relying on intermediaries based in Burkina Faso for recruitment. Mostly rural and from Mossi origin,[2] these boys are usually paid 50,000 CFA francs per month (around 81 US$), but only receive their salary when heading home, usually after a period of six months of work.[3] Many of them, however, do not receive any salary. They describe their work as part of the gift economy, hoping that their hard work will lead their boss – who is often a relative – to reciprocate by supporting them in their future endeavors.

In Guinea, by contrast, mining migration does not seem to involve intermediaries. Women and men, young and old, alone or with their families, travel freely to gold mining areas and have easy access to gold mining or related activities. At the sites where gold exploitation is based on manual techniques, people organize in work teams in which gains are shared equally. At the same time, it should be noted that not all family members (children, younger brothers or sisters, elderly people, and sometimes wives) have a say in the decision to become engaged in gold mining. These people accompany their decision-making member of the family on his travels and, once on site, participate directly or indirectly in the work of the mine.

An archipelago of mining sites

Mobility connects separate gold mining sites into a network: through mobility, they become part of an archipelago interconnected by various flows of information, people, and goods. Gold miners can easily obtain information concerning the different sites: location, systems of resource access, techniques, gold content, etc. Information and rumors about the discovery of new deposits, about productivity levels, or about the closing of sites promptly circulate between miners.

Since the 2010s, governments in both countries have been conducting military missions to expel miners and destroy mining camps. This has led to an intensification of mobility, described as "expulsion-induced migration" (Bolay 2018). Traders and service providers also move from site to site, in the footsteps of miners. The ASGM sector is not limited to young men digging pits with a pickaxe. There are also women who extract ore, older people who work in its transport or processing, and people who use a metal detector or a jackhammer, or who work in a mechanized crushing workshop. In addition, many other actors are involved

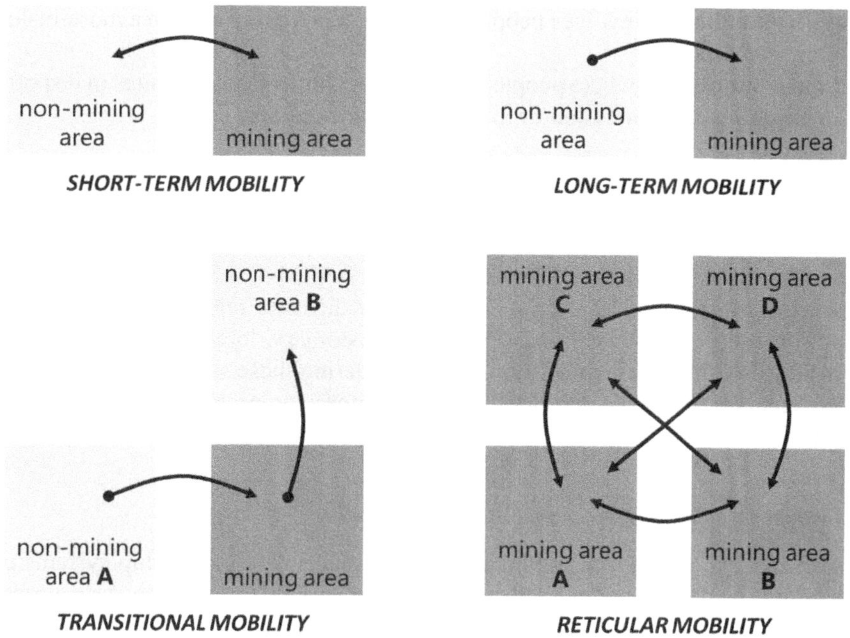

Figure 7.2 Forms of mining mobility in Guinea and Côte d'Ivoire. Figure by authors.

in the gold mining economy: gold buyers, shopkeepers supplying special equipment, blacksmiths, restaurant owners, laundrywomen, prostitutes, etc. In Guinea and Côte d'Ivoire, the mining migrations, whether they are induced by expulsion or undertaken voluntarily in the search of a better gold deposit, involve hundreds of thousands of people and connect different scales: local, national, and international.

These mobilities have contributed to the circulation of knowledge between countries. As an example, the logging techniques necessary for the exploitation of deep gold veins seem to have been developed in Burkina Faso. This innovation was then carried to Guinea and Côte d'Ivoire by gold miners who came from or had worked in Burkina Faso. In recent years, mobility between sites has facilitated the diffusion of ore extraction and processing machines (metal detectors, crushers, mills). Mobility has also contributed to the increased uptake of ore processing techniques based on mercury amalgamation or cyanidation. These techniques were previously unknown in Guinea and Côte d'Ivoire.

To summarize the above discussion, mining mobility in the ASGM sector in Guinea and Côte d'Ivoire since the 2000s has taken four main forms (as shown in Figure 7.2):

Short-term mobility occurs when people move to mining areas and sites for a few months to a few years but retain their place of origin as the main residence.

Transitional mobility occurs when people move to a mining area in order to accumulate enough capital to enable a long-term move to another, non-mining, area.

Long-term mobility involves people relocating to a mining area on a more-or-less permanent basis.

Reticular mobility is where people move between diverse mining sites in response to changing conditions and/or in pursuit of new opportunities.

Territorial governance of mining mobility

While a significant portion of gold miners come from non-mining regions, the inhabitants of gold-mining areas are also active in ASGM development, including by supervising mobility. This section concerns the different regulations and types of authority exercised over mining activities – customary, local, and national – and how they are linked to mining mobility. Considering these spatially fixed systems of governance regulating ASGM allows for a more comprehensive understanding of artisanal mining activities.

Individual vs. collective governance of mining mobilities

In West Africa, customary authorities and customary land ownership systems often form a parallel set of rules that are more-or-less integrated into the political-administrative system of the state. They locally organize access to land and, more generally, to income.

In Côte d'Ivoire, the governance of ASGM is not based on collective forms of local governance, but on exclusive access to land. Landowners directly manage activities on their plots, autonomously defining modalities of extraction and revenue sharing. This more fragmented governance is largely explained by the importance of the land issue in Côte d'Ivoire, a country marked by a plantation economy since colonization, and a generalization of individual modes of land appropriation through private property (Soro and Colin 2008; Colin 2021). Thus, the customary landowning lineages represented by one or more family members or the landowners themselves authorize, control, organize, and tax the activity performed on "their land," even though the activity is actually illegal. This more fragmented governance at the local level gives rise to the emergence of economic elites based on land rent. For example, at one of the sites studied, a local family happened to own the majority of the land on which the ore is extracted. In addition, given the increasingly widespread practice of subdivision and the possibility of buying land, a non-native of the village can take control of gold exploitation without having to submit to a centralized village institution.

In contrast, in the Upper Guinea region, "tutelage" relationships are the primary mechanism for organizing access to land and resources: migrants place themselves under the patronage of a local representative (Chauveau 2007). However, in the case of localities historically involved in ASGM – such as Kintinian, Doko, or Kourémalé (which are located in the Bouré and Sèkè regions) – a specific local institution, known as the *tombolomaw*, manages the organization of the activity. This organization does not question customary land ownership. Mining exploitation is carried out within the framework of a temporary delegation of administration rights to the *tombolomaw* and of use rights to the miners. In the historical regions of production, a landowning lineage cannot oppose the mining of its parcel but may receive compensation for the exploitation of its land (with that compensation managed by the *tombolomaw*).

Many studies refer to these *tombolomaw* as local organizations, similar to self-defense groups (Grätz 2007), which organize access to the site, ensure a form of moral justice, and centralize the taxation of activities (Lanzano and Arnaldi di Balme 2017). These entities also typically maintain close relations with the national security and defense forces, which contributes to the pluralization of security (Dessertine and Noûs 2021). The organizational and operational modalities of these collectives have been analyzed elsewhere, including in neighboring countries such as Mali (Panella 2007). Without going into more detail here, it can be observed that their presence testifies to the locally collective and grounded nature of the management of ASGM activities at the village level in Guinea. Even though the *tombolomaw* is not officially recognized by the Guinean mining code, state authorities (notably the prefectural mining directorates and the national security and defense forces) do not question their legitimacy nor their taxation system, and sometimes even benefit from it. The taxes imposed by the *tombolomaw* on gold miners (which vary from site to site, either as a share of bags on the pits or as a percentage of cash profits) can sometimes contribute to local development through the construction of infrastructures, such as schools, and in some cases even the establishment of a gendarmerie, raising the question of the intermingling of local and state authorities. In some areas, there is also a parallel and more individual option for inhabitants of mining localities: they can become associated with one or more wells as *nyado*, and as such do not participate in the work but can support the miners through small donations and receive a rent on the extraction following codified rules (Lanzano and Arnaldi di Balme 2017).

Just as the historical pattern of mobilities throughout this region has enabled the development of ASGM, local governance is also necessary for this activity. The case of the recent development of gold mining in northwestern Guinea illustrates the importance of the local regulation of mining mobility. The Gaoual region is not a historical mining area. Gold was discovered there at the beginning of 2021 by a local resident who had worked for a long time in mining sites in the northeast of the country. The spread of information about the presence of gold resulted in an inflow of tens of thousands of people within a few days, including gold miners from the historical mining areas. The number of new arrivals quickly exceeded the local population. Faced with these population movements, the regional authorities quickly decided to ban gold mining and deployed several military and police units. The gold miners were tracked down, some were arrested and had their equipment seized, but the exploitation did not stop. Gradually, the defense and security forces on the ground began to directly negotiate with the gold miners. Outside any official regulatory framework, the military started to give and withdraw exploitation authorizations. In this way, the local population was deprived of the mining income but suffered from the effects of the gold boom. Gold mining continued for several months in a climate of violence between gold miners, local people, and the military.

At the end of 2021, the situation changed. The management of the mines was transferred to the local authorities. An alliance was formed between representatives of the municipalities and mobile gold miners. This resulted in the establishment of *tombolomaw* made up of locals with little experience of gold mining and mobile gold miners. As in the historical mining regions, the *tombolomaw* were in charge of the governance and taxation of gold mining. They were recognized as legitimate by the miners, even though this institution had no historical legitimacy in this area.

The different ways that gold mining is controlled in our case study areas thus reflect both local mining histories (or lack thereof) and local-level power relationships. While in Guinea, the governance of ASGM is organized by specialized centralized organizations at the village level, in Côte d'Ivoire, individualized access to land determines control over gold mining. This admittedly schematic comparison allows us to emphasize the importance of locally anchored modalities of control and management of ASGM activities.

Negotiating foreignness to become a local

In both countries, local gold mining governance is controlled by people from the mining area. At first glance, mobile gold miners, who constitute an important, and in some cases a major, part of the labor force, are excluded from governance. However, in the case of Guinea, the foreign gold miners – non-native of the mining area whether they are Guinean or not – do not remain passive in the face of local regulation but participate in its negotiation and enactment.

Governance varies between sites in Guinea, depending on the intensity of the activity, the amount of gold in the deposit, and the local history of ASGM. In the prospecting phases or at the beginning of exploitation, taxes collected by the *tombolomaw* are very low or even non-existent. Once they are established, their level increases as exploitation develops. Conversely, when exploitation decreases, mobile gold miners often negotiate a tax reduction with the *tombolomaw*, and the possibility of easily moving to other mining sites contributes to their capacity to do so, as being mobile gives them leverage in interactions with territorial authorities. As a result, and despite some variation, tax rates are relatively similar across the different sites in Upper Guinea.

Furthermore, the indigenous/foreigner division is not strict. On arrival, all non-natives to the locality are considered foreigners, regardless of their origin or nationality, and thus excluded from mining governance. But settling in a mining locality over a long period of time or being recognized as an experienced gold miner can lead to integration into the governance of gold mining: by these means, some mobile gold miners become *tombolomaw*. Certain actions facilitate this negotiation of foreignness, in particular marrying a person from the mining locality or building a house and setting up residence in this locality. Moreover, in some localities in Upper Guinea, foreign lineages have gradually been integrated into local customary authorities. In Kintinian, for example, in the 2010s, 20 non-indigenous lineages were integrated into the council of elders, which is responsible for appointing the *tombolomaw*. The possibility to negotiate foreignness seems nonetheless limited to nationals.

In Côte d'Ivoire, integration of mobile gold miners into the governance of mining is also limited to nationals, because the regulation of mining activity is linked to land ownership, which has been legally inaccessible to non-nationals since 1998. But even if some locals are involved in gold mining and are often self-employed, most landowners delegate management to intermediaries who may be non-native to the village or foreigners, and who often are those who actually invest in the activity. The owner benefits from the activity through fixed taxes (on machines, transport, and access to the site) and the managers benefit from the profitability of the activity. These flows can sometimes merge if the manager works for the owner, who then accumulates both taxation revenue and profits. This system, while it does not explicitly allow for the integration of outsiders into local governance, nevertheless allows their enrichment, ensuring them an important decision-making position. For example, at one of the sites we studied, a non-native was regularly consulted by the local chieftaincy because of his key economic position in the gold sector.

Fixing miners through formalization

Since the structural adjustment policies promoted in the 1980s and 1990s by international financial institutions, mining policies in Guinea and Côte d'Ivoire (and more broadly in Africa) have been part of a cumulative process of liberalization that has aimed to create an attractive environment for private investment (Campbell 2004; Campbell and Hatcher 2019). These policies reduce the land accessible to ASGM and secure the most promising areas for industrial mining. Prospection and extraction permits are largely distributed to foreign industrial and semi-industrial companies, subsequently facilitating the violent eviction of gold miners (in Guinea, for example, see Bolay [2014] or Dessertine and Noûs [2021]), even when the companies have no actual activity on site. The distribution of these permits facilitates state and corporate control over vast territories which until then had largely escaped state authority, thus undermining customary land tenure and political authority, without having to challenge these local structures openly. Mostly justified through the paradigm of "modernization," these policies have been rationalized as a way to reduce the negative social and environmental effects of gold mining while securing a fiscal revenue. Indeed, central authorities had been largely excluded from gold mining taxation, although decentralized administrative structures had often set up forms of local and informal taxation. For example, in both Guinea and Côte d'Ivoire, the decentralized administrative offices of the countries' Ministries of Mines may levy annual taxes on small mechanized crushers, municipalities may levy other taxes on the transport of ore, and gendarmes may levy tolls on the passage of gold miners and ore transporters. These taxes are set up outside any legal framework or are implemented as fines. These practices contribute to the informalization of public governance, by creating a lucrative and highly personalized system of taxation that limits the interest of institutions to enforce national regulations.

The mining codes of 2013 for Guinea and 2014 for Côte d'Ivoire include artisanal activities and propose ways to formalize them, notably through artisanal mining authorizations. These apply to natives of the country or to legal entities whose capital is wholly (in the case of Guinea) or mainly (in Côte d'Ivoire) owned by nationals. However, as in other West African countries, gold miners do not take ownership of these mechanisms, and formalization is not effective. Moreover, formalization requires a mastery of bureaucratic processes, high expenditures over several years, and successful insertion into a political network of patronage.

Even though we have encountered several miners in Côte d'Ivoire who had willingly invested a lot of time and money in the hopes of formalizing their activities, none of them had yet succeeded. In addition, other legal frameworks are ineffective. For example, in Guinea, the state is setting up ASGM corridors in the "gaps" between industrial mining titles. These mining corridors are not based on potential gains for gold miners or on the local presence of gold mining and are thus largely unsuited. In addition, although technical ASGM management teams were supposed to be established at the prefectural level, only four prefectures have such a technical team, and only one seems to be operational.

Dispersing the miners through repression

In this context of difficult formalization, state policy toward informal gold mining is fluctuating and highly contingent. Long-standing tolerance can brutally turn to active repression as soon as state interests appear at risk. For example, in the case of the expansion of industrial mines on artisanal gold mining sites, security forces may be mobilized to expel gold miners who had previously been tolerated. This repressive pattern forces workers to be highly mobile from site to site, adapting to the volatile targeting of law enforcement.

These processes take several forms but always combine two dimensions of security: information and intervention. Indeed, at the sites, security forces – mainly gendarmes – are permanently stationed and cannot afford to be too repressive given the daily proximity they maintain with the gold miners and local inhabitants. They are known for "eating into the activity" – that is, recuperating in a variable and informal way a share of the rent, following a logic of corruption. But they are also the ones who centralize information, sometimes to the benefit of gold miners. This was the case, for example, in one of the sites in Côte d'Ivoire, where certain elements of the gendarmerie warn the miners when an eviction or equipment seizure operation is planned. Violent evictions are carried out by mixed brigades composed of police, gendarmes, and military personnel, created *ad hoc* in Guinea, or institutionalized in Côte d'Ivoire, where a special brigade for the repression of mining code violations (BRICM) was created in October 2018. Members of these brigades, who are not acquainted with the miners, are known to commit the most violent acts, including against individuals, as evidenced by the events of 5 October 2022, when at least five individuals were killed by the BRICM in the site of Kokumbo.

The reinforcement of an anchored territoriality through mechanized mining mobility

While in the early 2000s production increased through the intensification of old extraction practices, gold mining now uses a variety of more-or-less mechanized techniques, which has significantly increased the initial capital necessary to invest in the activity. Mechanization thus transforms mining mobility and gold mining governance.

Mechanization and (im)mobility

The mobility of workers is transformed by new extraction techniques that require a fixed location. Timbering practices, for example, make it possible to exploit deeper wells. This results in a longer period of well exploitation. The use of motor pumps and aerators also allows exploitation to continue during the rainy season. Conversely, the use of metal detectors results in a form of hyper-mobility (Bolay 2016; Dessertine 2016; Dessertine and Noûs 2021): with a minimal working depth, prospection is carried out over large areas through constant movement.

Revenue-sharing arrangements are also impacted by mechanization. In manual mining, gold miners are relatively independent. They freely join teams, usually gender-mixed, in which the sharing of earnings is equitable. In the case of mechanized gold mining, there is an economic division between capital and labor. Investors, in particular those who own the most expensive equipment (trucks or hydraulic shovels, which represent investments of tens of thousands of US$), usually are not directly active in gold mining sites. The work is performed by people who do not own the production tools. Conversely, the owners of crushing workshops and of mechanized sluices (which require a few hundred to a few thousand US$ worth of investment) are usually directly involved in the activity. They may have to negotiate the purchase and storage of ore or work themselves crushing ore. The level of capital invested is inversely proportional to the investors' proximity to the working site. Few gold miners have the capital to invest in machinery, and thus most of the miners in the mechanized forms of ASGM resign themselves to an employee status, while the owner of the capital bears both the risk of loss and possibility of gain.

Workers can be paid either in the form of a percentage of earnings or through a fixed wage. The development of wage forms tends to reduce mobility because wage earners receive a regular and predictable income. Unlike self-employed gold miners, they do not have the possibility of having "lucky strikes." The chance for a major gold find disappears. In this situation, there is no longer any interest in being involved long term in this sector, unless one can substantially invest in the activity. Given the hard work and the regularity of earnings, it is more likely for workers in mechanized workshops to carry out shorter periods of gold mining with a defined objective, for example, to finance an extra-mining activity or an investment in gold mining equipment for themselves.

Finally, the mechanization of gold mining results in fewer job opportunities. On the one hand, the amount of labor required to achieve the same level of production is reduced. On the other, while manual forms of gold mining present job opportunities to a wide range of people – from the most experienced and athletic men mining in galleries to the weakest reprocessing the ore after the first wash – mechanized gold mining is mostly performed by physically strong young men, or by people with financial capital or particular skills (drivers, conveyors, logisticians, etc.).

This results in a masculinization of the activity and, more generally, in the exclusion of certain categories of people, in particular the elderly, women, and people without any financial capital.

A disputed local governance in Guinea

Existing taxation systems are not always adapted to these new forms of extraction. In Upper Guinea, the governance of the *tombolomaw* is based on the exploitation of resources through pits and galleries at sites where mining activity is artisanal and spatially concentrated. The new forms of gold mining based on mechanized equipment that have developed over the last 20 years are not, or only to a limited extent, included in the *tombolomaw* management system, weakening its governance. As we have seen, metal detector prospecting leads to an expansion of activities over larger areas, thus circumventing the spatial control of the *tombolomaw*. In the gaps in the mining governance system arising from mechanization, new actors position themselves at the local level. Public actors (communes, districts, youth offices, decentralized administrative services) and private actors (landowners) seek to gain control over the regulation of mechanized artisanal activities in order to capture mining rents. With the more widespread use of crushers and mechanized washing ramps, which require a relatively fixed place, the landowning lineages play an increasingly key role in the governance of ASGM – for instance, by organizing mechanized extraction on their plots themselves or by requiring a share of bags or cash income – thus leading to an individualization of mining rents. This evolution reinforces the issue of land ownership in a context where part of the land was managed at the village level without the necessity of individual appropriation (Arnaldi di Balme and Lanzano 2013; Dessertine 2019).

However, statements about the crumbling power of the *tombolomaw* in the context of mechanization must be tempered. Because mechanized forms of gold mining do not replace manual practices but are developed in parallel, the new actors involved in the regulation of gold mining do not always replace the *tombolomaw*. This institution is sometimes even recreated, as in the case of Gaoual discussed above. Even when circuits of redistribution are reconfigured, there is a structural continuity that lies in the hierarchical relation between mobile and immobile actors of ASGM. To put it differently, the objective ultimately remains the same – to control mining mobility and to reap the benefits locally – but the place taken by the different actors is negotiated according to the availability of resources, the spatial and technical organization of exploitation, and the local power structures.

For example, in the Kintinian locality in northeastern Guinea, where artisanal open-pit mining sites have emerged over the past decade, the extracted ore is transported by truck to mechanized sluice workshops installed near rivers. Collection of taxes on the transport of ore by the *tombolomaw* was taken over by the city council. In 2022, for several months, the municipal authorities had to deal with strong criticism and contestation over the management of this income. In this context of tension between the mayor's office and the other actors, the customary authorities decided that the management of the tax on trucks would be transferred from the mayor's office to the district office. Within three months, this institution redistributed more money to the customary authorities than the city hall had given them previously. At the time of writing, local taxation of ore transport is still controlled by the district office.

Nonetheless, landowners appear to be playing an increasingly key role in the governance of ASGM in Guinea, by directly organizing mechanized extraction on their plots. This individualized access to mining rent could supplant the hierarchical tutelage relationships that prevailed in customary law. This has already had a direct impact on the landscape, including the agricultural landscape, by leading to forms of occupation that aim to establish control over a delimited space, as shown by an increase in plantations to secure land ownership. According to national law, the establishment of a plantation allows the recognition of the private character of the juridical property and entitles the owner to a rent on exploitation. The exploitation of a parcel by a mining company also allows the landowner to receive compensation, reinforcing the trend of creating plantations to secure land ownership (Mbodj 2011).

Building on land ownership in Côte d'Ivoire

The rise in the power of landowners and associated changes to property geographies have led to a transformation of production relations in gold mining in Côte d'Ivoire. A distinction is no longer made between locals who benefit from mining rent and foreign workers, but between landowners and workers. Whereas the *tombolomaw* retain some power in ASGM governance in Guinea, landowners are the key actors in Côte d'Ivoire, where the plantation economy had already individualized land ownership. But even though some landowners may invest in the activity, according to our observations and interviews, the most important investors are non-natives. Here, we witness the emergence of entrepreneurs within the ASGM sector who compensate for the absence of landownership by creating mutually profitable alliances with native landowners. This is the case of one of them, born in the north of Côte d'Ivoire, who was employed in the 1980s in an industrial mine and then in an artisanal cooperative in one of our study sites. As this cooperative did not survive, he used his technical skills to invest in informal semi-mechanized mining in the early 2010s, just as the gold boom was beginning in the region. He quickly became rich and decided to build his household in the village. He and his sister are now pivotal to the informal exploitation of the site, particularly because they are also in charge of marketing the ore.

We also observed the possibility for investors to play with the boundary between formal and informal domains, by obtaining titles to control access to land, for example, especially through mining permits. In one of our sites of investigation, an artisanal authorization permit has become a formal screen for informal exploitation: the gold miners benefit from the protection of the permit holder under an agreement to sell their ore directly to the latter at a discount. The permit holder does not need to invest in equipment because he allows the gold miners to continue their activity – a subsistence activity for many of them – and to benefit directly from their exploitation by positioning himself as a commercial intermediary. The multiplication of these entrepreneurs, with one foot in the informal sector and the other in the formal sector, testifies to the hybridity and ambiguity of the governance systems of ASGM.

Conclusion

This comparative perspective on the techniques of exploitation, the organization of work, and the political regulation of ASGM in Guinea and Côte d'Ivoire offers an original vantage point for the analysis of ASGM governance.

In both countries, in a context of inadequate national regulation, local institutions enact, enforce, and monitor rules that govern artisanal mining at the local level. This control consists of supervising the activity, ensuring the security of the sites, and allowing the local population to benefit from a rent on the mines. This involves a triple form of governance: (i) territorial governance as it is controlled by local actors acting in their own territory; (ii) resource governance since the rules and institutions are specific to mining; and (iii) governance of mobile persons since the objective of these rules is to regulate the exploitation of the resource by persons who are often from outside the mining territory.

In both countries, customary governance always includes possibilities to negotiate foreignness, and access to resources is never restricted by nationality. Whoever one is, it is possible to work on the mining sites as long as local rules are respected and taxes are paid. In contrast, official regulatory texts of Guinea and Côte d'Ivoire are based on a clear definition of the foreigner as a non-national. The possibility of formalizing gold mining then favors nationals, regardless of their region of origin, which differs from the native/non-native division currently used at local level.

In both countries, the governance of gold mines is linked to local architectures of power. In Guinea, governance of non-mechanized mines is managed collectively at the village level by strong institutions recognized by gold miners. The increasing mechanization of mining activities is a new context in which other actors, notably landowners, are seeking to participate in the governance of mines. This has contributed to the emergence of a legal debate over land tenure regulations. In Côte d'Ivoire, by contrast, mining activities have developed in a context in which the land issue was already politically and socially central. In Côte d'Ivoire, the structuring of the activity around private land ownership seems, on the one hand, to have reinforced dynamics of unilateral land appropriation, favoring masculine power (as it is men who predominantly own land) and excluding non-nationals (as they are legally barred from land ownership). On the other hand, the emergence in rural Côte d'Ivoire of successful trajectories of capitalist mining entrepreneurs, deprived of land ownership, indicates that the growing importance of financial capital in the organization of the activity can also open new opportunities to women and non-natives.

Despite a rapid reconfiguration of local, national, and international power relations around gold mining due to the growing mechanization of informal mines, this comparative and temporal perspective reveals a certain continuity: mobility is essential to the extraction and processing of gold ore, and territorial governance – in its various forms – aims to control this mobility in order to enable and benefit from mining exploitation. This situation differs from the gold rushes in Saharan-Sahelian regions, where ASGM can give rise to new localities or revitalize existing ones (Gagnol, Magrin, and Chevrillon-Guibert 2020) and where mobility produces a more ad hoc governance, with probably more potential for conflict than in the regions we have studied. In Guinea and Côte d'Ivoire, on the other hand, regulation of mobility is completely integrated into local governance and in the production of rural spaces (Dessertine 2021). The development of ASGM is therefore both framed and enabled by this governance framework that reconciles two dynamics that are, in fact, more complementary than opposed: mobility and territorial fixity.

Notes

1 According to the World Bank, in 2021, the rate of population growth in Côte d'Ivoire was 2.5 percent, and in Guinea it was 2.4 percent.
2 The Mossi are a West African people living primarily in central Burkina Faso, northern Ghana, Togo, and Benin.
3 1 US$ = 612 CFA francs in May 2023. With a salary of 50,000 CFA francs (around 81 US$) per month, they receive the equivalent of 490 US$ for six months of hard work.

References

Arnaldi di Balme, Luigi, and Cristiano Lanzano. 2013. "'Entrepreneurs de la frontière': Le rôle des comptoirs privés dans les sites d'extraction artisanale de l'or au Burkina Faso." *Politique Africaine* 131 (3): 27–49. doi: 10.3917/polaf.131.0027.

Arrault, Jean-Baptiste. 2005. "Du toponyme au concept? Usages et significations du terme archipel en géographie et dans les sciences sociales." *L'Espace Géographique* 34 (4): 315–328. doi: 10.3917/eg.344.0315.

Blundo, Giorgio, and Pierre-Yves Le Meur. 2009. "Introduction: An Anthropology of Everyday Governance Collective Service Delivery and Subject-Making." In *The Governance of Daily Life in Africa: Ethnographic Explorations of Public and Collective Services*, edited by Giorgio Blundo and Pierre-Yves Le Meur, 1–37. African Social Studies Series 19. Leiden: Brill.

Bolay, Matthieu. 2014. "When Miners Become 'Foreigners': Competing Categorizations within Gold Mining Spaces in Guinea." *Resources Policy* 40: 117–127. doi: 10.1016/j.resourpol.2014.02.001.

Bolay, Matthieu. 2016. "'Il faut être là où l'or sort !' De l'itinérance temporaire au maintien d'un mode de vie mobile chez les orpailleurs de Haute Guinée." *Stichproben* 16 (30): 111–135. doi: 10.25365/phaidra.266_06.

Bolay, Matthieu. 2017. "Gold Journeys: Expulsion-Induced Mobility and the Making of Artisanal Mining Spaces in West Africa. An Ethnography of Itinerant Labour at the Bottom of the Gold Supply Chain." PhD diss., Neuchâtel University.

Bolay, Matthieu. 2018. "Reversed Rushes? Expulsion as a Dominant Feature of Gold Miners' Mobility in Guinea and Mali." *Resource Worlds*, February 25.

Bolay, Matthieu. 2022. "Miners on the Move: Expellable Labor, Flexible Intimacies, and Institutional Recast in West Afrixan Artisinal Gold Mines." *American Ethnologist* 49 (4): 521–535. doi: 10.1111/amet.13108.

Boyer, Florence. 2013. "La figure de Pénélope ou l'immobilité dans le contexte des migrations circulaires." *e-Migrinter* 11: 53–65. doi: 10.4000/e-migrinter.229.

Brugger, Fritz, and Jessica Zanetti. 2020. "'In My Village, Everyone Uses the Tractor': Gold Mining, Agriculture and Social Transformation in Rural Burkina Faso." *The Extractive Industries and Society* 7 (3): 940–953. doi: 10.1016/j.exis.2020.06.003.

Campbell, Bonnie, ed. 2004. *Enjeux des nouvelles réglementations minières en Afrique.* Uppsala: Nordiska Afrikainstitutet.

Campbell, Bonnie, and Pascale Hatcher. 2019. "Neoliberal Reform, Contestation and Relations of Power in Mining: Observations from Guinea and Mongolia." *The Extractive Industries and Society* 6 (3): 642–653. doi: 10.1016/j.exis.2019.06.010.

Champy, Muriel. 2015. "Ni enfants, ni adultes: Une lecture comparative de la 'jeunesse' (Burkina Faso)." *Ateliers d'anthropologie* 42. doi: 10.4000/ateliers.10024.

Champy, Muriel. 2022. *Faire sa jeunesse dans les rues de ouagadougou*. Sociétés Africaines 25. Nanterre: Société d'ethnologie.

Chauveau, Jean-Pierre. 1979. *Notes sur l'histoire économique et sociale de la région de Kokumbo, Baoulé-Sud, Côte d'Ivoire*. Travaux et Documents de l'O.R.S.T.O.M 104. Paris: O.R.S.T.O.M.

Chauveau, Jean-Pierre. 2007. "Transferts fonciers et relation de 'tutorat' en Afrique de l'Ouest: évolutions et enjeux actuels d'une institution agraire coutumière." *Journal des Sciences Sociales* 4: 7–32.

Chevrillon-Guibert, Raphaëlle, Laurent Gagnol, and Géraud Magrin. 2019. "Les ruées vers l'or au Sahara et au nord du Sahel. Ferment de crise ou stabilisateur ?" *Hérodote* 1 (172): 193–215. doi: 10.3917/her.172.0193.

Colin, Jean-Philippe. 2021. "La marchandisation de l'accès à la terre dans des contextes ruraux ouest-africains." *Le Mouvement Social* 4 (277): 117–132. doi: 10.3917/lms1.277.0117.

Cresswell, Tim. 2010. "Towards a Politics of Mobility." *Environment and Planning D: Society and Space* 28 (1): 17–31. doi: 10.1068/d1140.

De Lange, Albertine. 2007. "Child Labour Migration and Trafficking in Rural Burkina Faso." *International Migration* 45 (2): 147–167. doi: 10.1111/j.1468–2435.2007.00407.x.

Dessertine, Anna. 2013. "Le lu ne meurt jamais : mobilités des individus et pérennité de la résidence dans un village malinké de Guinée." *Géocarrefour* 88 (2): 131–138. doi: 10.4000/geocarrefour.9058.

Dessertine, Anna. 2016. "From Pickaxes to Metal Detectors: Gold Mining Mobility and Space in Upper Guinea, Guinea Conakry." *The Extractive Industries and Society* 3 (2): 435–441. doi: 10.1016/j.exis.2016.02.010.

Dessertine, Anna. 2019. "Une justice foncièrement autre?" *Revue Internationale des Études du Développement* 238 (2). Éditions de la Sorbonne: 141–164. doi: 10.3917/ried.238.0141.

Dessertine, Anna. 2021. *A quoi tient le village. Espaces et mobilités en pays malinké (Guinée)*. Nanterre: Société d'Ethnologie.

Dessertine, Anna, and Camille Noûs. 2021. "Hybrid Territorialisation: A Reconfiguration of Rural Spaces through Gold Mining in Upper Guinea." *Political Geography* 86: 102348. doi: 10.1016/j.polgeo.2021.102348.

Gagnol, Laurent, Géraud Magrin, and Raphaëlle Chevrillon-Guibert. 2020. "Chami, ville nouvelle et ville de l'or. Une trajectoire urbaine insolite en Mauritanie." *L'Espace Politique*. 38. doi: 10.4000/espacepolitique.6562.

Grätz, Tilo. 2003. "Les chercheurs d'or et la construction d'identités de migrants en Afrique de l'Ouest." Translated by Roland Marchal. *Politique Africaine* 91 (3): 155–169. doi: 10.3917/polaf.091.0155.

Grätz, Tilo. 2004. "Les frontières de l'orpaillage en Afrique occidentale." *Autrepart* 30 (2): 135–150. doi: 10.3917/autr.030.0135.

Grätz, Tilo. 2007. "Vigilante Groups and the State in West Africa." In *Order and Disorder: Anthropological Perspectives*, edited by Keebet von Benda-Beckmann and Fernanda Pirie, 74–89. New York: Berghahn Books.

Grégoire, Emmanuel, and Laurent Gagnol. 2017. "Ruées vers l'or au Sahara: l'orpaillage dans le désert du Ténéré et le massif de l'Aïr (Niger)." *EchoGéo*. doi: 10.4000/echogeo.14933.

Hannam, Kevin, Mimi Sheller, and John Urry. 2006. "Mobilities, Immobilities and Moorings." *Mobilities* 1 (1): 1–22. doi: 10.1080/17450100500489189.

Hilson, Gavin. 2016. "Farming, Small-Scale Mining and Rural Livelihoods in Sub-Saharan Africa: A Critical Overview." *The Extractive Industries and Society* 3 (2): 547–563. doi: 10.1016/j.exis.2016.02.003.

Hilson, Gavin, and Chris Garforth. 2012. "'Agricultural Poverty' and the Expansion of Artisanal Mining in Sub-Saharan Africa: Experiences from Southwest Mali and Southeast Ghana." *Population Research and Policy Review* 31 (3): 435–464. doi: 10.1007/s11113-012-9229-6.

Huntington, Heather, and Kate Marple-Cantrell. 2022. "The Importance of Artisanal and Small-Scale Mining for Rural Economies: Livelihood Diversification, Dependence, and Heterogeneity in Rural Guinea." *Journal of Rural Studies* 94: 177–191. doi: 10.1016/j.jrurstud.2022.06.004.

Kopytoff, Igor. 1987. *The African Frontier: The Reproduction of Traditional African Societies*. Bloomington: Indiana University Press.

Lahiri-Dutt, Kuntala, ed. 2018. *Between the Plough and the Pick: Informal, Artisanal and Small-Scale Mining in the Contemporary World*. Acton: Australian National University Press.

Lanzano, Cristiano, and Luigi Arnaldi di Balme. 2017. "Des 'puits burkinabè' en Haute Guinée: processus et enjeux de la circulation de savoirs techniques dans le secteur minier artisanal." *Autrepart* 82 (2): 87–108. doi: 10.3917/autr.082.0087.

Mbodj, Faty Bineta. 2011. "Boom aurifère à l'est du Sénégal, l'ouest du Mali et au nord-est de la Guinée : mutations socio-économiques et spatiales d'anciennes marges géographiques et économiques." PhD diss., Université Gaston Berger de Saint-Louis du Sénégal, Université Paris 1 Panthéon-Sorbonne.

Ngom, Ndeye Marame, Modou Mbaye, David Baratoux, Lenka Baratoux, Kouassi Ernest Ahoussi, Jean Kan Kouame, Gayane Faye, and El Hadji Sow. 2022. "Recent Expansion of Artisanal Gold Mining along the Bandama River (Côte d'Ivoire)." *International Journal of Applied Earth Observation and Geoinformation* 112. doi: 10.1016/j.jag.2022.102873.

OECD. 2018. *Gold at the Crossroads: Assessment of the Supply Chains of Gold Produced in Burkina Faso, Mali and Niger*. Paris: OECD.

Ouédraogo, Lala. 2019. "Orpaillage artisanal et développement rural." PhD diss., Laval University.

Panella, Cristiana. 2007. "L'éthique sociale du damansen: Éducation familiale et orpaillage artisanal dans le Basidibé (Wasolon, Mali)." *Cahiers d'Études Africaines* 47 (186): 345–370. doi: 10.4000/etudesafricaines.7261.

Pijpers, Robert. 2014. "Crops and Carats: Exploring the Interconnectedness of Mining and Agriculture in Sub-Saharan Africa." *Futures* 62: 32–39. doi: 10.1016/j.futures.2014.01.012.

Soro, Debegnoun Marcelline, and Jean-Philippe Colin. 2008. "Marchandisation, individualisation et gestion intra-familiale des droits sur la terre en basse Côte-d'Ivoire." Économie rurale 1–3 (303–305): 154–168. doi: 10.4000/economierurale.657.

Thorsen, Dorte. 2006. "Child Migrants in Transit: Strategies to Become Adult in Rural Burkina Faso." In *Navigating Youth, Generating Adulthood. Social Becoming in an African Context*, edited by Catrine Christiansen, Mats Utas, and Henrik E. Vigh, 88–114. Uppsala: Nordiska Afrikainstitutet.

Section III

Melanesia

8 Mining-induced in-migration in Papua New Guinea

Glenn Banks and Tobias Schwörer

Introduction

In this chapter, we examine human migration in the context of four large-scale mining projects in Papua New Guinea (PNG). By considering migration patterns at projects that are at different stages in their life cycles, we aim to illustrate the spatial, temporal, social, economic, and environmental dimensions of migration. Our focus is on migration that is (mostly) indirectly induced by mining projects, rather than the more conventional, narrow focus on in-migration of employees to mining projects, though we recognize that at times this distinction can be blurred.

Drawing on fieldwork undertaken by ourselves and others, as well as previously published work by Bainton and Banks (2018), we document and analyze population movements at mining projects not only at different scales but also in distinct socio-environmental settings. We begin with two large-scale exploration prospects, the Frieda River copper mine and the Wafi-Golpu copper mine. While both are in the pre-approval phase, they have already induced significant and complex flows of migration over several decades. We then turn to the large-scale Porgera and Lihir gold mines, which have been operating for close to three decades. The very different migration patterns around these two projects are shaped by geographic, economic, and social factors specific to these settings. What all these projects have in common is a regulatory and institutional setting in which the vast majority of the land around these projects is held under communal tenure by local "landowning communities." As will become apparent, not only is the term "landowner" problematic and contested, but the legal status of land around these projects also shapes how projects develop, and the forms of in-migration that occur.

Our primary aim here is to refine and update a schematic model developed by Bainton and Banks (2018) that attempts to capture the intersections and feedback loops between migration, social processes, and landscapes that emerge when mining interests arise, exploration gets underway, and mines become operational in specific locations. Regional context, the specific forms of local social organization, community responses, and mining company policies are additional elements that shape these flows, as we address in the discussion. By focusing on two different stages in the project life cycle (advanced exploration and mature operational

DOI: 10.4324/9781003313236-11

stages) across four projects, we are able to demonstrate the variety of population movements that occur as projects are proposed and developed, and then as they expand and extend over decades.

The chapter proceeds from here by outlining key elements of the model developed by Bainton and Banks (re-elaborated here as Figure 8.1), highlighting how it responds to the Melanesian context in which land is an essential element for the construction of social relations. Three additional components that we introduce, and we contend need to be accounted for, are: (i) the broader regional context within which mineral projects occur, (ii) the variety of ways in which local conceptions of kinship or relatedness are deployed and shape processes of in-migration, and (iii) the agency of resident communities as well as policies by the state and mining companies that actively seek to constrain, modify, or in some cases enhance movements of people into project areas. We then briefly introduce and discuss migration processes and flows around each of the case study projects. All four projects have an extended and complex social, political, economic, and exploration history and sit within a complicated ethnographic and geographic setting. We can barely scratch the surface of all these aspects. Instead, we highlight some of the distinctive characteristics of migration processes and flows around each that can help illuminate aspects of the broader model. After the case studies, we draw on the insights gained from each to suggest refinements to the conceptual model developed by Bainton and Banks (2018).

A basic schematic on mining and migration in Melanesia

Land and social relationships – and specifically *social relationships to land* – sit at the heart of society in the hundreds of distinct Melanesian communities.[1] In essence, our argument is that migration – induced and driven by mining projects, though not limited to the movements of employees – fundamentally reshapes the relationships between peoples and land, as well as social relationships between people. In this sense, as Bainton and Banks (2018, 458) contend, land acts as

> more than simply a site of tradition, or a simple commodity, but instead as a site for the intersection of processes of identity, modernity and relationships, a place where the assemblages and affordances around land come together to produce often highly charged contests over meanings, identity and development.

Figure 8.1, re-created from Bainton and Banks (2018), provides a basic graphic of the connections between mining, in-migration, and development in Melanesian settings. As Bainton and Banks (2018) acknowledge, "real life" processes are much more dynamic than this simple model suggests, but this is a useful way of representing the key elements that need to be considered when analyzing in-migration processes. The key factor that initiates in-migration is landscape and environmental change, which commences from the exploration stage (starting with

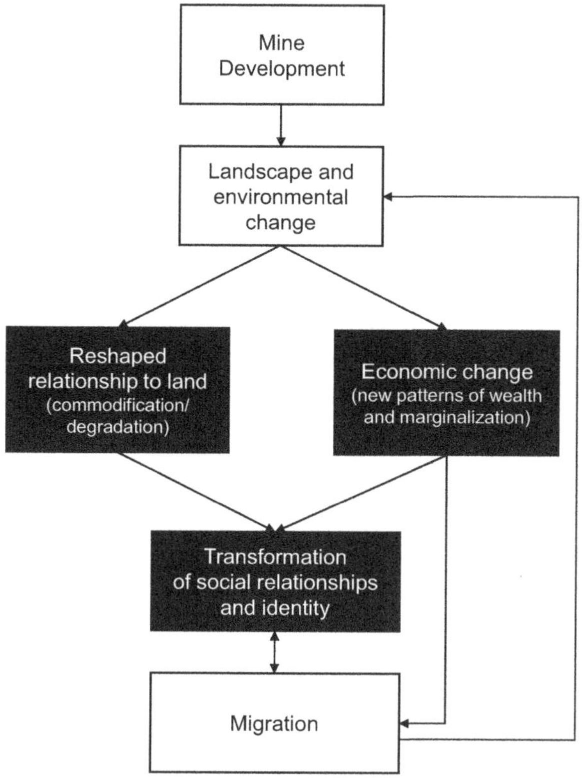

Figure 8.1 A schematic of the relationships between mining and in-migration. Elaborated by Marilyn Ishikawa; based on Bainton and Banks (2018, 454).

the demarcation of lease areas) and increases dramatically during the construction and operational phases of a mine. These physical-environmental changes, in turn, are entangled with the commodification of land and stratification of local society.

Commodification is the process whereby land acquires a new exchange value (in addition to its traditional use values). It is sparked by landowners receiving cash compensation for land damaged, disturbed, destroyed, or (partially) alienated and – if the land is part of a mining lease – royalties, lease payments, and potentially dividends from landowner equity holdings in the mines (Banks 2003). This represents a fundamental reconfiguration of society-environment relations for Melanesian communities (Bainton and Macintyre 2013; Jacka 2015; Schwoerer 2023). Another significant consequence, as Bainton and Banks (2018, 454) observe, is that "[o]ver time, group definition is rearticulated according to the ability of individuals and lineage groups to claim customary ownership over impacted land and therefore gain access to compensation and mining benefits."

The rapid development of new patterns of wealth and marginalization in turn leads to greater stratification of local society (Bainton and Banks 2018; Beer and

Schwoerer 2022). These processes begin long before the initiation of extraction, with the expectations of wealth that can be derived from the mining development, culminating in a re-ordering of social relations and the development of unequal and hierarchical factions among customary land claimants. Whichever group can most convincingly lay claim to the land inside a mining lease is sure to reap most of the benefits from mining. This process is usually guided from the top by a few mostly well-educated and well-connected men from the local landowning communities, who then set up and attempt to retain control over landowning associations and companies, through which most of the benefits of the mining project are distributed. The process of determining landowning communities also means that certain groups will be sidelined, be it because their claims are not acknowledged, or because their claims are much smaller than those of similar groups. This can create stark inequalities within a regional area, wherein a few landowning communities receive large compensation packages and royalties, and other groups receive much less or nothing at all, while the latter simultaneously bear a share – sometimes even a disproportionate share – of negative ecological and social impacts. Within landowning groups, similar exclusionary processes take place: politically influential men rise to the top of these new groups and associations formed, while women and younger men are often sidelined. Those at the top of the hierarchy are then in a perfect position to benefit from the distribution of compensation payments and royalties. It is this wealth generated by the mine and accumulated at the top that then triggers in-flows of people not actually working at the mine or in mine-related businesses.

Migration in PNG generally, whether it concerns rural-urban migration or migration to mining sites, is often patterned and shaped by pre-existing or newly established relations of kinship and wider notions of relatedness. Kin networks are crucial links that enable people to move to new places, as these are underpinned by notions of solidarity, generosity, and reciprocity. Especially in formerly rural areas that now host mining projects, kinship is often the main determining factor in gaining rights of residence, access to land for subsistence, or gainful employment. As we will show, in-migration to mining sites is occurring through the activation of links of descent, or via marriage relations, whereby the marriage of a member of one kin group (often a woman) to a member of a mining community enables chain-migration of other members to follow.

While kin relations are important, there are also wider forms of relatedness and non-kin-based networks of solidarity that play a role in migration, especially the pan-Melanesian notion of *wantok* (Tok Pisin for "one talk," i.e., same language). Originally designating people from one's language group who band together in places far away from home, the term nowadays is also loosely applied to friends and acquaintances who exhibit a degree of sameness in terms of place or even just province of origin, or in terms of schooling, religion, or work experience. *Wantok* relations, premised on notions of mutuality and solidarity, are thus both informal self-help networks in a new, unfamiliar environment, as well as a new form of relatedness resembling kinship in a mostly urban (or resource extraction) setting (Schram 2014).

There is a temporal element to these forms of migration and the effects on identity and relationships, which the model in Figure 8.1 does not really capture. As Bainton and Banks observe (2018, 445), once processes of in-migration are activated, they can become self-perpetuating; or, as they put it: "As mining continues and the number of migrants increases, this expands the possibilities for other more opportunistic migrants, whose connections to the 'original' landowners may be increasingly tenuous, even conflictual." This comes through in the cases that follow.

Case studies

Each of the four projects discussed below has a specific geographic and historical context, of which, in the interest of brevity, we can only sketch the most pertinent aspects. We focus on some of the ways in which mining-induced migration reflects and shapes the processes we have outlined in the previous section. Notably, with the partial exception of Lihir, none of the projects have or are planned to have, a stand-alone, isolated "mining town," as there is at Ok Tedi and Panguna on Bougainville in PNG, as well as many other locations globally. (See Figure 8.2 for locations of four case study sites.)

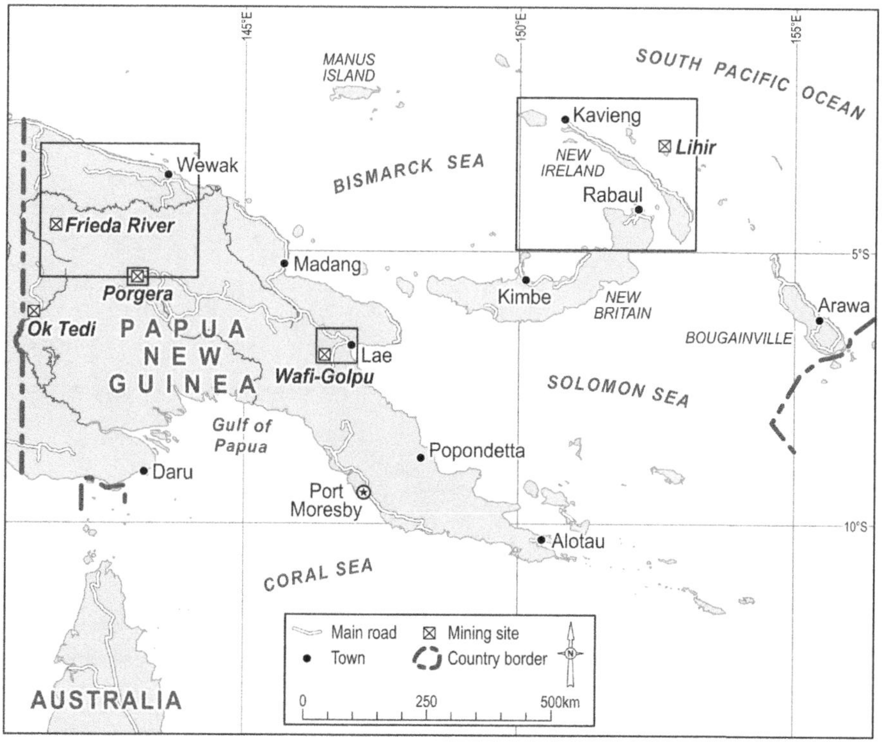

Figure 8.2 Papua New Guinea, showing locations of the case study sites. Map by Tim Nolan (Blackant Mapping Solutions) and authors.

Frieda River

The Frieda River project is currently owned by PanAust, an Australian-based subsidiary of Guangdong Rising Holding Group Co., Ltd. It has gone through numerous development phases and is currently progressing through the final planning and permitting stage, having been a site of exploration on and off for more than 50 years. The large-scale copper and gold deposits are located on the northern slopes of the central mountain ranges of PNG, within the upper catchment of the Frieda River (see Figure 8.3). Located in Sandaun (West Sepik) Province, the land constituting the site drains into the Sepik River. There is no road access to the site, which is currently heavily dependent on river transport up the 500-km-long Sepik River, though the project would be connected by 70 km of proposed new road from the Sepik to the mine. The latest iteration of the project would see an integrated tailings disposal and hydroelectric dam and an extensive road connection from the project site to the coast at Vanimo, making the project a significant regional infrastructure project. It would completely alter the current geographical isolation of the site. This isolation, to date, has meant there has been no large-scale in-migration of people coming from outside of the project area and its immediately surrounding communities.

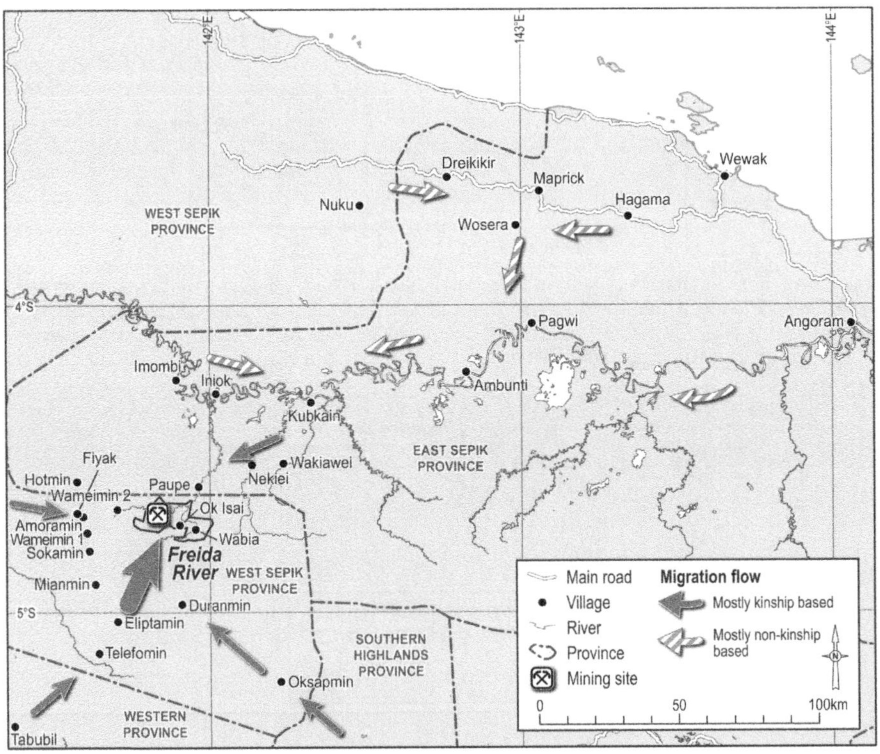

Figure 8.3 Frieda River project area, showing likely major migration pathways. Map by Tim Nolan (Blackant Mapping Solutions) and authors.

The Frieda River complex sits at the intersection of at least four language and ethnic groups, which together constitute a mixture of groups from the more mountainous south (the Ok/Telefol) and the lowland/hinterland groups (Bainton and Banks 2018, 452; Skrzypek 2021). The broader socio-geographical context is one in which mobility and migration have been an important part of social and cultural life – and central to livelihood practices – for centuries. Each of the main groups – especially the Telefomin, Miyanten, and Paupe – have traditionally accepted outsiders. For instance, the inclusion of women and children from defeated enemy groups historically has been utilized to bolster the strength of the group; within the Sepik groups, traditional obligations have extended to peoples from allied totemic groups. In addition to these formal kinship-based connections, many people in the area have additional relationships, varying in strength and nature, with people distant from the project area. For example, the Telefomin view the Mountain Ok speakers across the border in Indonesian Papua as kin, and some Indonesian Mountain Ok have already worked at the Frieda River project.

Importantly, the social organization of the different groups has informed past migration to the project area and will likely do so in the future. The potentially affected communities along and near the Sepik River have strong patrilineal clan systems, where descent and rights flow through the father's line. In contrast, the Telefomin and Miyanten have descent systems that are close to being cognatic, whereby connections can be claimed through both the father's and mother's lineage. This automatically means that each individual from these groups (especially if male) can activate potential links with a wide range of landowners and other residents in the project area. Polygamy is a common practice in most of the communities around the project area, and as is the case around other mining projects, there is already evidence that most of the additional wives of mining area landowners originate from outside the project area. This has the additional effect of facilitating in-migration of the wife's relatives, and indeed in-migration has often been an implicit part of bride price settlements.

The current setting reflects a century of trade, warfare, and inter-marriage that has generated flows of people and relationships and produced "dramatic shifts in settlement and population" (Jorgensen 2007, 61) prior to the colonial period. More recent migration into and between the mine-area communities (in anticipation of mining and associated revenue streams and benefits) has reshaped what was already a weakly delineated and dynamic social landscape.

By the mid-1990s, multiple overlapping claims of ownership were being made by four different language groups and across the proposed mining area, new settlement locations had been formed, and a growth in alluvial mining in the wider area was also attracting new migrants from surrounding areas. (According to current plans, several existing alluvial gold areas will disappear beneath the waters of the proposed hydropower dam, but other presently less frequented pockets of alluvial gold will escape flooding and will continue to attract migrants to the area.) As described by Bainton and Banks (2018, 458),

> Loosely defined boundaries and flexible forms of descent reckoning facilitated the movement into the area of a significant number of people over the period. This has led to a complex and socially heterogeneous set of

settlements whose inhabitants have a far more diverse set of connections and origins than in the past.

Until the late 1970s, the mountain people adjacent to the project were mostly involved in limited, traditionally driven, forms of migration. Since then, however, there has been widespread involvement in and movement toward the Ok Tedi mining project and its mining township of Tabubil. Thousands of Telefol and Oksapmin have moved either as workers or simply as squatters to the areas around Tabubil and have become used to residing near or sometimes working for mines. There is a widespread expectation that when Frieda enters the next phase of development, many of those currently in the Tabubil area will "return to their own project."

Together, these factors have created conditions which have facilitated and will encourage further in-migration when (rather than if) the project proceeds to construction and operation. The major migration pathways to the project area, shown on the map in Figure 8.3, will be from Telefomin along kinship lines, and from the Sepik River area. The latter in-migration may not be as numerous as that from Telefomin, but its impact will be even more pronounced. In-migration from the Middle Sepik and Wosera-Gawi is already occurring, largely feeding through Ambunti, making this a key strategic position for these migration pathways. Tabubil, the Ok Tedi township, is another anticipated source of migrants, particularly for people from the related Oksapmin and Telefomin groups. Smaller numbers of migrants have already appeared from related Mountain Ok-speaking groups from the Indonesian side of the border, and even non-kin traders from Indonesia.

Major flows of people from other parts of the country are unlikely, but there is a possibility of some migrants coming from areas around Kiunga in the Middle Fly (Western Province), Porgera, and Enga Province (and by proximity and residence in Porgera, people from Hela and the Southern Highlands), and from Madang Province. As most of the kinship connections between these groups and the project area landowners are tenuous at best, these migrants are likely to be more opportunistic and entrepreneurial and would work to build local connections – such as business partnerships with locals – once residing in the project area.

The Frieda River project demonstrates how precolonial migration patterns within a broad regional complex came to be built on by the movement of people into project-area communities over an extended 50-year period of large-scale mining exploration. As the various exploration projects have generated activity and expectations of wealth, people have moved into the broader area, typically drawing on notions of belonging and connection to one (or sometimes more) of the recognized "landowning" groups. This has created a complex and highly charged social landscape, even before mine construction has occurred (Banks and Jackson 2011).

Wafi-Golpu

Wafi-Golpu is a large-scale copper and gold deposit. It is currently held in a 50:50 joint venture between the Denver-based Newmont Gold (which took over the Australia-based Newcrest Mining in mid-2023) and South Africa-based Harmony

Gold Mining Company Limited. It is in the mountainous interior of Morobe Province, relatively close to the historic Wau-Bulolo mining district. Like Frieda River, the Wafi-Golpu project is progressing toward the final planning and permitting stages. It is the first large-scale underground mine in PNG, and as such will likely have a much smaller operational footprint than Frieda River or other existing mines (Bainton and Banks 2018, 452). In addition, the proposed block cave mining operations will be heavily automated, leading to a relatively small workforce of approximately 850 full-time employees (compared with 2,500 in Porgera, for example).

The proximity of the project to the major port city of Lae (only 70 km away; see map in Figure 8.4) adds what Bainton and Banks (2018, 452) refer to as "another level of social and geographic complexity." The current exploration camp is supplied through a road branching off from the Lae-Bulolo Highway. However, a new access road will connect the mine with the Highlands Highway, a critical arterial road that runs roughly 700 km through the densely populated highlands region down to Lae, and which will doubtlessly encourage in-migration. The access road to Wafi-Golpu will branch off from the Highlands Highway around the 43-mile marker out of Lae, near Dzifasing, cross the Markham and Watut Rivers, and continue up the Watut Valley. Also, a 100-km-long infrastructure corridor – with a concentrate, tailings, and fuel pipeline – will connect the mine to the port in Lae. This corridor follows the Watut and Markham Valleys, traverses customary and state land, and thus generates impact communities along its path.

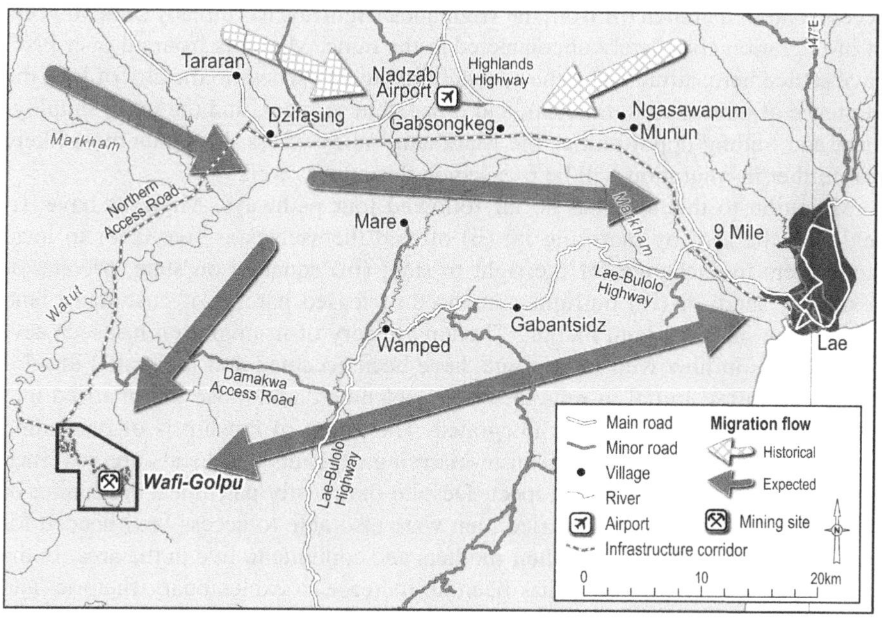

Figure 8.4 Wafi-Golpu project area, showing major existing and expected migration pathways. Map by Tim Nolan (Blackant Mapping Solutions) and authors.

In the area immediately around Wafi-Golpu, local population movements have fluctuated with opportunities for employment in exploration activities since the deposit was discovered in 1977. Various intersecting factors have produced a situation in which no one group can claim exclusive rights to the project area. Long-standing land disputes over the Wafi-Golpu project area between the immediately adjacent Babuaf, Yanta, and Hengambu groups and more distant groups, like the Piu and Wampar (the latter of which claim to be the original landowners in the faraway past, but then moved away), will influence negotiations with the company over land access and the final outcomes of the benefits-sharing agreements (Church 2022). One of the consequences has been to generate a steady flow of work for mediators, lawyers, and lands specialists (Bainton and Banks 2018, 456). However, in-migration from outside the immediate region so far has been rather negligible, as the area is relatively remote and economic opportunities are still few.

Once construction begins, the principal focus of in-migration might not necessarily be the immediate area around the mine. The mining company Wafi-Golpu Joint Venture has set up its headquarters and is renting a large residential estate, as staff housing, at 9 Mile, a suburb of Lae, where the Wau-Bulolo Highway branches off from the Highlands Highway. From there, it shuttles personnel to the exploration site, and will likely continue to do so even when construction gets underway. Subcontractors working for the mining prospect have so far also mostly set up their businesses between Lae and Nadzab airport, in an uninterrupted stretch of suburbs, informal settlements, and former villages that have grown far beyond their traditional boundaries. This area, from Lae all the way up to Dzifasing, where the new access road will branch off from the Highlands Highway, has already seen 50 years of in-migration completely unconnected to the mine. Migrants from all over PNG have settled here, attracted by the accessibility and closeness to the city of Lae, the existence of educational, medical, and other state services, and the many employment and trading opportunities the Markham Valley offers. It is thus most likely that further in-migration will be focused on this area as well.

Migration to this area has so far followed four pathways. Migrants have: (i) settled in the area by marrying in; (ii) offered themselves as free labor to local landowners in exchange for the right to stay; (iii) squatted on state, private, or customary land; or (iv) outright purchased or leased parcels of customary land through the informal land market. The long history of in-migration has seen several changes in how well in-migrants have been accepted and integrated into local communities. Initial in-migrants were welcomed, and those that married into the communities were quickly integrated. The idiom of kinship is of overriding importance, and relations between in-marrying migrants and locals are governed by norms of reciprocity and respect. Despite the mostly patrilineal reckoning of descent, the children of in-married men were also able to access land needed for subsistence activities through their mothers and continue to live in the area. In the last 15 years, however, there has been an increase in exclusionary rhetoric, and some children of in-married men, for example, have no longer been allowed to plant more permanent cash crops like cocoa. Also, daughters have been informed that they could no longer remain in the community if they also chose to marry non-local men. The notion of patrilineal descent is thus increasingly being used to corral the local community, although the extent to which exclusion happens is still very much dependent on social relations. Most of these children, for example, still figure on the lists drawn up to form landowners' associations and incorporated land groups (Bacalzo 2012, 2019; Bacalzo, Beer, and Schwoerer 2014).

The second pathway for in-migration is through relations of patronage and clientelism. This pathway is mostly used by people from the more remote and mountainous parts of Morobe and Madang Provinces, especially from the Rai Coast and Menyamya. They have for decades migrated to the Markham Valley to offer themselves as free labor to local landowners looking for help in medium- to large-scale agricultural projects, such as cattle ranches, large-scale chicken farms, cocoa orchards, or peanut and watermelon fields. The increasing population has also led to a situation where subsistence gardens are increasingly susceptible to theft, and hence some migrants have been given the role of watchmen. In all these cases, there is an informal agreement that such migrants can build houses and subsistence gardens on the land of their hosts in exchange for free labor. There is quite a large social divide between these in-migrants and local communities, and the relation is one of dependency and deference. The migrants are vulnerable to exploitation and eviction, and in the past, whole groups have been chased off in cases of conflicts with local landowners.

The third and fourth pathways have been through squatting on state, private, or customary land, or leasing or buying customary land from local landowners. While squatting on state and private land can entail a certain vulnerability to eviction, the topic of eviction has been a contentious political issue, and any large-scale eviction exercises have needed to use a certain degree of force. A much more secure option for in-migrants is thus buying or leasing land from customary landowners. Even though customary land cannot legally be sold to individuals, all major cities in PNG have experienced the development of an informal market for customary land (Koczberski et al. 2017; Rooney 2017). The sale of land has been contentious among the local landowning communities, especially when some members of the landowning groups sell land (and pocket the money) while others are opposed to this practice. This dynamic has at times led to a "tragedy-of-the-commons"-type situation in which land has been sold to pre-empt others from selling it. Evicting these migrants has been difficult, as they invariably request the return of their payment, which has often already been spent on consumption, or invested into education or small- to medium-scale businesses. The rampant sale of land to people unconnected to these communities (especially in the area between Lae and Nadzab airport) has led to a situation in which local communities are now increasingly dwarfed by the large numbers of migrants living on their land. This has contributed to numerous social problems and conflicts between locals and migrants, who are no longer as easily integrated into the communities as they were 50 years ago (Beer 2008, 2017). Notably, however, two communities in the Markham Valley, namely Dzifasing and Tararan, so far have not sold any land (Schwoerer 2023).

In a pattern at odds with the other examples in this chapter, there is the potential for a continuous flow of people and things between the Wafi-Golpu mine, the nearby urban center of Lae, and the suburban zone stretching along the Highlands Highway (see Bainton and Banks 2018, 458). Migrants attracted by opportunities arising from the mine are just as likely to move to settlements in and around Lae as they are to move to the mining site itself. This is different from the typical image of rural-to-resource migration flows and is more akin to a pattern of circular flows, as discussed elsewhere in this volume. The landowner representatives of Hengambu, Yanta, and Babuaf, for example, are already well connected to Lae and usually have a *pied-a-terre* in one of the settlements or suburbs surrounding the city.

The mining company is aware of the potential for large-scale immigration into the Watut Valley upon completion of the new access road from the Highlands Highway and has already undertaken some measures planned to limit immigration. This includes setting up a major cocoa project along the Watut Valley to give the communities alternative means of livelihood unconnected to the mine, so that people value the land not just for the potential compensation it can bring. This transformation of formerly unused land into cocoa orchards also ensures that there is less space available for potential settlements to form. The same reason – "to fill up the land to prevent settlements of in-migrants attracted by the mine" – was also given by the former Huon Gulf Member of Parliament for why he supported an oil palm project in the area around Dzifasing and Tararan (where the new mining access road will connect to the Highlands Highway). Whether such measures will have any success is difficult to predict, however.

Porgera

The Porgera gold mine is located in the western portion of Enga Province in the highlands of PNG, at the end of the Highlands Highway (see Figure 8.5). Construction of the mine began in 1989, and by 1992, it was the third largest gold-producing operation in the world. The mine was closed in 2020 because of the PNG Government's refusal to renew the operation's Special Mining Lease. However, as of the time of writing, it is on track to reopen with a new PNG-majority ownership structure, and Barrick Gold maintaining its operator status.

Two factors have been central to the nature and extent of subsequent mine-induced migration: kinship and geography. Prior to the development of the mine, the main inhabitants of the Porgera Valley had been the Ipili people, who were a relatively small and marginal population. Before European contact, they had sought connections – through marriage, exchange, and trade – with the much larger Engan population to the east, and with the equally large Huli population to the south. The Ipili employed an open and inclusionary form of cognatic descent reckoning, meaning that individuals are potentially able to claim a connection to group belonging, rights to land, and significant social relationships following three generations on either parent's side (Biersack 1996; Golub 2014). With the development of the new mine in the western portion of the valley, "the surrounding populations drew on these pre-existing social linkages to try to access the massive revenue flows and new infrastructure that was being developed" (Bainton and Banks 2018, 452).

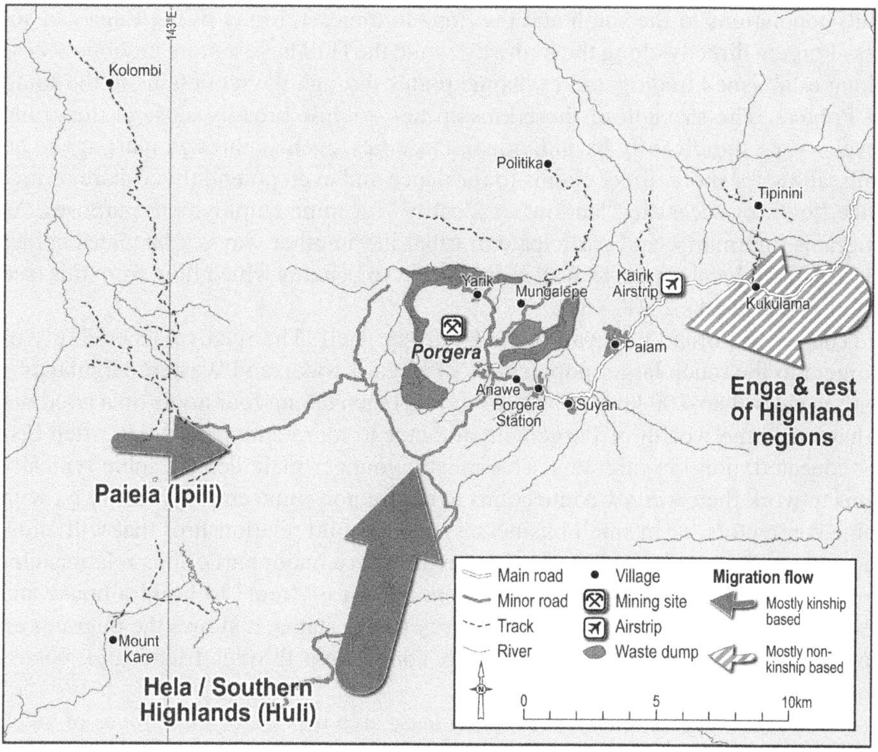

Figure 8.5 Porgera gold mine area, Enga Province, showing major migration pathways. Map by Tim Nolan (Blackant Mapping Solutions) and authors.

In terms of the geography, the Highlands Highway has, since the 1980s at least, provided direct road access to Porgera for the extensive population of the Highlands region (stretching all the way down to Lae and Madang). It is also the "end of the road," and the wealth and business activities associated with the mining project sit amid a large area of relatively low levels of development and economic activity: there are, in other words, few reasons to leave once people arrive at Porgera.

The construction and initial operation of the mine was the start of significant and transformative migration that has created a variegated social mosaic in the Porgera Valley. The district's population has grown from around 5,000 in 1980 (prior to mining development) to 10,000 in 1990 (at the start of the mine operation; see Banks 1997) and ballooning to over 45,000 by 2012 (Bainton and Banks 2018). It is worth noting that these figures include the expatriate and national "fly-in/fly-out" mine workforce, but this is a relatively small component: involving less than a thousand workers at any one point in time.

There are broadly two major pathways of migration to Porgera. As already noted, the first, and initially the dominant form, has been kinship-based in-migration from surrounding areas. This includes the immediate Ipili neighbors in the Paiela region, who held close kinship ties with Porgeran Ipili, as well as the more distant

Huli populations to the south and the Enga to the east. Many of the Enga can access Porgera directly along the highway, while the Huli have a more arduous access along established trading and exchange routes through the mountains to the south of Porgera. The strength of these kinship ties – while broadly close at the group level – vary significantly by individual: close ties, such as through marriage to an Ipili, allow for more direct claims to residence and even potentially a share of revenue flows, or access to "landowner identity" for mine employment purposes. As migrants intermarry, and participate in tribal life in other ways, it becomes harder to deny them local status, as well as the access to benefits which flow from this (see Bainton and Banks 2018, 455).

The second broad pathway is the Highway itself. The relative accessibility of Porgera to the much larger populations of Enga, Jiwaka, and Western Highlands – with the less-than-200 km road trip from Mt Hagen taking four hours on a good run – has made the wealth of Porgera an attractor to more entrepreneurial, often better educated, non-kin migrants. This mostly younger male demographic typically aims to work their *wantok* connections to secure non-mine employment (e.g., with mine contractors, or in small businesses) and to build relationships that will allow them to reside and stay at Porgera. Money is often a major part of this relationship-building: employees may pay landowners a form of "rent" to build a house and reside on their land. While there is precarity to this status, it allows the migrants an opportunity to build stronger, more secure connections through friendships, contributions to exchange, and even marriage.

Within the Porgera Valley, the mining lease area was the original focus of much kin-based migration, for the reasons discussed above. There was even movement of Ipili from the lower parts of the valley – an area of long-standing alluvial gold mining – into the upper valley where the mine and its leases were concentrated (Banks 1999). From the mid-1990s, small-scale mining activity – initially of mine tailings, then of ore that locals extracted from both the open pit and even from waste dumps – drew significant numbers of younger, mostly male migrants into the communities immediately around mine operations. This created flashpoints of conflict with mine security, police, and other security forces, and between the different social groupings in the area. Burton (2014) and Jacka (2019) have documented the rise and intensity of "tribal fights" in and around the mine area. The recent deaths, in 2022, associated with the national election at Porgera were also tied to longer-standing frictions between groups and individuals.

Management of this in-migration by the state or the corporation – both of whom would like to reduce the risks of social disintegration that such migration can feed – is difficult at best. As at Lihir (see below), local representatives insisted on local preference clauses in the original agreements negotiated at the start of the mine. These were tools that, in theory at least, would reduce migration for employment, by ensuring that Porgerans were given preference for positions at the mine, subject to experience and skills. However, as time has gone on, adherence to these clauses has become more difficult as a second generation of "locals" has pursued lucrative mine employment, and pressure has increased on those delegated to decide on the "local-ness" of applicants. Where the mine has attempted to remove "migrants" from within its lease area – often with the support of the original landowners – there have been conflicts and repeated reports of human rights abuses (HRW 2010). Consistent with the feedback loop shown in Figure 8.1, continuing in-migration at Porgera has generated additional pressures on the local environment and increased local conflict, with migrants often being held to blame. (For a more detailed analysis of the social changes occurring at Porgera, see Bainton and Banks 2018.)

Lihir

The Lihir gold mine is located in the caldera of an extinct volcano on the main island of Aniolam in the Lihir Group of Islands, off the east coast of New Ireland (see Figure 8.6). In 1995, the developer, the landowners, and the government negotiated an integrated benefits package that enabled the project to proceed. Mining operations commenced two years later, in 1997. The mine is now owned and operated by Newcrest Mining Limited (recently acquired by Newmont Mining, as noted above). Operations are expected to continue beyond 2035. According to Bainton (2010), large-scale mining development on the island has led to deep social divisions and

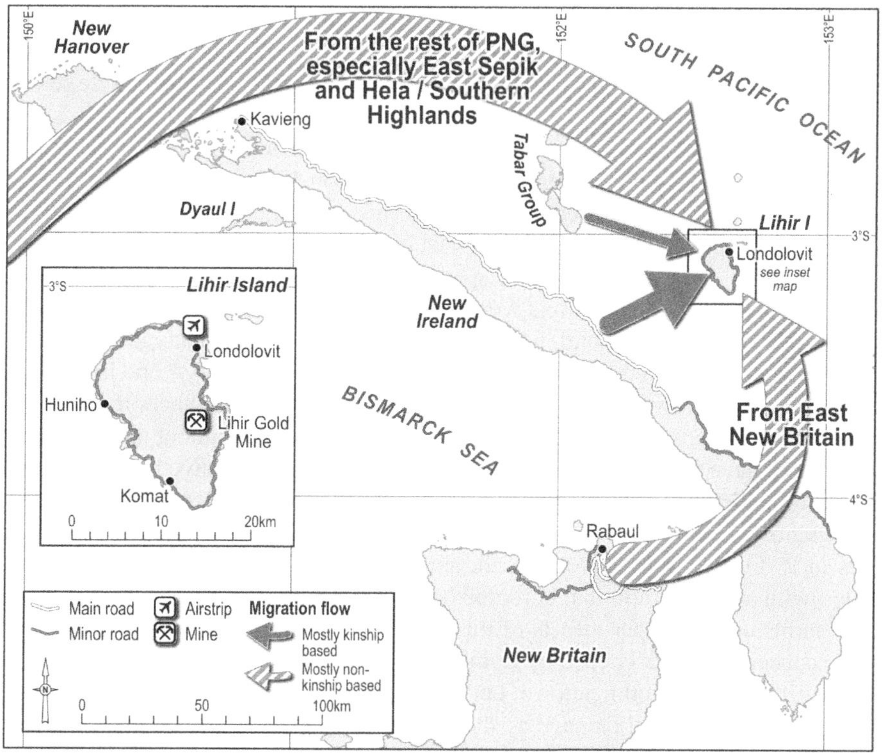

Figure 8.6 Lihir gold mine, New Ireland Province, showing major migration pathways. Map by Tim Nolan (Blackant Mapping Solutions) and authors.

complex disputes among different groups centering on the distribution and control of mine-related benefits. The scale of in-migration at Lihir has not matched levels observed at Porgera or those predicted for the Wafi-Golpu or Frieda River projects (Banks and Jackson 2011). Nonetheless, the island setting's contained geography has amplified and intensified the impacts of these inward flows (Bainton 2017).

When the mine started, in 1996, the population of the Lihir group of islands was around 9,900. Just a few years earlier, the non-Lihirian population on the island was estimated to be around 300 (Filer and Jackson 1989). At the establishment of the mine, and with the support of the local population, a sophisticated population register (the Village Population System, or VPS) developed by Drs John Burton and John Vail was put in place and for two decades resulted in the best record of demographic change and analysis for any of the Melanesian mining operations. According to the VPS, by 2010, there were more than 8,000 migrants and another 3,500 Newcrest employees in camps residing on the island, alongside just over 15,000 Lihirians.

Initially, the primary pathway for migration was via the ocean. The vast majority of the initial migrants arrived by small boat, along kinship lines from the nearby islands and the New Ireland mainland, and into communities where these kin-related migrants were able to assert claims to residence (even if these claims were at times contested). Over time, and facilitated by regular flights to Lihir from Port Moresby and other island sites, people unconnected by kinship to Lihir have increasingly been moving to Lihir. By 2010, while New Ireland was still the largest single source of migrants, these flows were almost matched (for migrants on the mining leases at least) by those from East New Britain, East Sepik, and Southern Highlands, and 14 of PNG's provinces were represented among the non-Lihirians on the company's leases. Police charge book records mirrored this pattern, with people from these same four provinces accounting for the vast bulk of charges in 2009 and 2010.

There was – and is – significant spatial variation in the locations where non-Lihirians took up residence. Most were concentrated around the three wards on the north-eastern side of the island, where the mine and the most directly affected communities are located. In these wards, Lihirians were in the minority – and in one case were outnumbered by non-Lihirians by more than two to one. Others have moved closer to the mine. As at Porgera, landowners have not been without agency and in some cases have facilitated migrant flows. For example, there have been numerous reports of landowners renting out their new mining-derived modern material houses to migrants, while the landlords returned to their bush material houses. As with other sites in PNG, and indeed globally, the migrants were dominated by working-age males, with a gender ratio that bordered on four males to one female.

In terms of the deeper effects of this migration, and specifically with reference to the conceptual model posed by Bainton and Banks (2018) – and re-elaborated in Figure 8.1 – the commodification of land occurred rapidly. This has extended to using claims of environmental damage by the mine to support demands for compensation (Bainton and Macintyre 2013; Macintyre and Foale 2004). Over time, group definition has been rearticulated according to the ability of individuals and lineage groups to claim customary ownership over impacted land and therefore gain access to compensation and mining benefits. The increasing numbers of kin-related migrants who have moved into these communities, combined with the other changes large-scale mining has wrought, have made the local claim-making landscape much more complicated and changed local power dynamics and kinship structures in significant – and not always welcome – ways (see Bainton and Banks 2018, 458).

Attempts by the mining company to manage migration have included proposed schemes to move migrants from within lease boundaries (around 800 in 2011) by providing air tickets "home" and cash payment. However, these plans have not been carried out, at least to date. Despite some overt xenophobia and aggression from Lihirian leaders and the local government toward "migrants" (and evidence that the migrant community in 2009–2010 was the source of a majority of crimes), there has been a continuing reluctance – coupled with an inability – to pursue concrete measures to remove or reduce the number of migrants on the island.

Discussion and conclusions

The case studies portrayed above present four significantly different geographies across PNG: an isolated island (Lihir); a remote valley currently unconnected by road (Frieda River); Porgera "at the end of the road" (Highlands Highway); and a site in close proximity to a major highway and urban center (Wafi-Golpu). These regional geographic and developmental contexts have had a significant influence on the ways in which in-migration occurs. Not surprisingly, road access to the mining site from large parts of mainland PNG increases in-migration significantly, even more so if the surrounding landscape is otherwise little developed. An important element that is most apparent from the Wafi-Golpu case is that mining-induced migration cannot be regarded as distinct from pre-existing traditional and modern regional migration flows: historical and regional context shapes contemporary mining-induced migration to a significant degree. The whole Markham Valley, from Lae up to Dzifasing, has seen already a large influx of migrants from elsewhere in PNG for more than 50 years completely unconnected to the mining project. These people were attracted by the location close to the city of Lae, the many state services available, and the employment and trading opportunities in the Markham Valley. Most of the impact of further mining-related in-migration will therefore likely also take place in those areas (and less in the immediate area of the mine), which stands in contrast to the other sites.

The four cases have also shown that the notion of kinship, or wider forms of relatedness, such as those captured by the notion of *wantok*, are often defining pathways in accepting, allowing, and integrating in-migrants into local communities, although notions of descent can also be employed to restrict immigration. The four cases demonstrate markedly different notions of kinship and descent. As we have shown in the case of Frieda and Porgera, loose groupings and flexible reckonings of mostly cognatic descent are clearly encouraging in-migration, by offering pathways to become kin through marriage and descent. Once integrated into local communities, the distinction between "local landowners" and "outsider in-migrants" becomes increasingly difficult to uphold, particularly for the next generation. Unilineal reckonings of descent, on the other hand, as in the case of Lihir and Wafi-Golpu, are not necessarily a hindrance to in-migration and integration into local communities, but they can be rhetorically employed to restrict in-migration. With increased tensions around the migration issue, such arguments can also be invoked as grounds for limiting the number of recipients of royalties and compensation payments.

Also apparent in various sites are the agentic practices of individuals and communities in mining project areas, who are typically very aware of the potential of in-migration and its effects, and over time may work to prevent or at least strategically reduce it. One of the key aspects of limiting or restricting immigrants are reckonings of descent, which themselves can become contested over time, as the Porgera case in particular shows. Who counts as "local" when it comes to assigning preferential employment status, for example, is a difficult question, especially when different forms and notions of descent and relatedness clash.

Further developing the schematic model of Bainton and Banks (2018), we have argued that broader regional geographic, economic, and political contexts constitute an important backdrop to mining-focused in-migration, shaping the nature and form of population flows. Further, the specific forms of social organization within the communities around these projects shape the internal processes – around identity, representation, and conflict, for example – that both encourage or constrain in-migration and mitigate or attenuate its effects.

As the case studies presented here demonstrate, in-migration pathways, patterns, and motivations are not easy to predict or manage and can be a source of frustration and conflict for multiple parties. In the PNG context, mining corporations, the state, and even local communities have at times tried to devise policies and practices to limit in-migration and mitigate the impacts. However, these efforts have mostly been unsuccessful, as the cases of Porgera and (to a lesser extent) Lihir demonstrate. Here, we concur with the argument put forward in Bainton and Banks (2018) that these failures are largely the result of companies failing to invest in the requisite social management capabilities (Filer, Banks, and Burton 2008; Kemp and Owen 2018; Owen and Kemp 2017) that would help them to better understand and navigate the social, geographical, and political complexities of mining localities.[2] The outcome is that mine-induced in-migration is, for the most part, unregulated and uncontrolled, and looms large as a source of friction, driving multiple crises of identity, of social conflict, and of environmental pressures. Hence despite the concerns and efforts of communities and corporations to shape and mitigate movements of people to these mining sites, this in-migration sits at the heart of the processes that Filer (1990), in the context of the Panguna copper mine, famously labeled as "social disintegration."

Notes

1 There are more than 800 distinct recognized language and cultural groupings in Papua New Guinea, and across these, there are many different forms of social organization and cultural practices, including ways of connecting with and inheriting land, for example. The wide variety of customary land tenure forms found across Papua New Guinea

reflects the cultural diversity of the region and the range of ways that different social groups organize themselves and reckon descent.

2 As Bainton and Banks (2018) also note, the reluctance of PNG governments to become directly involved, due to the political stakes, has been another constraining factor, although this is not easily addressed.

References

Bacalzo, Doris. 2012. "Transformations in Kinship, Land Rights and Social Boundaries among the Wampar in Papua New Guinea and the Generative Agency of Children of Interethnic Marriages." *Childhood* 19 (3): 332–345.

Bacalzo, Doris. 2019. "Economic Changes and the Unequal Lives of Young People among the Wampar in Papua New Guinea." In *Pacific Youth: Local and Global Futures*, edited by Helen Lee, 57–78. Canberra: ANU Press.

Bacalzo, Doris, Bettina Beer, and Tobias Schwoerer. 2014. "Mining Narratives, the Revival of «Clans» and Other Changes in Wampar Social Imaginaries: A Case Study from Papua New Guinea." *Journal de la Société des Océanistes* 138 (1): 63–76.

Bainton, Nicholas A. 2010. *The Lihir Destiny: Cultural Responses to Mining in Melanesia*. Canberra: ANU Press.

Bainton, Nicholas A. 2017. "Migrants, Labourers and Landowners at the Lihir Gold Mine, Papua New Guinea." In *Large-Scale Mines and Local-level Politics: Between New Caledonia and Papua New Guinea,* edited by Colin Filer and Pierre-Yves Le Meur, 313–351. Canberra: ANU Press.

Bainton, Nicholas A., and Martha Macintyre. 2013. "'My Land, My Work': Business Development and Large-Scale Mining in Papua New Guinea." In *Engaging with Capitalism: Cases from Oceania*, Vol. 33, edited by Fiona Mccormack and Kate Barclay, 139–165. Bingley: Emerald Group Publishing Limited.

Bainton, Nicholas A., and Glenn Banks. 2018. "Land and Access: A Framework for Analysing Mining, Migration and Development in Melanesia." *Sustainable Development* 26 (5): 450–460.

Banks, Glenn. 1997. "Mountain of Desire: Mining Company and Indigenous Community at the Porgera Gold Mine, Papua New Guinea." PhD diss., The Australian National University.

Banks, Glenn. 1999. "Gardens and Wantoks." In *Dilemmas of Development: The Social and Economic Impact of the Porgera Gold Mine 1989-1994*, edited by Colin Filer, 160–190. Canberra and Port Moresby: Asia-Pacific Press, (Pacific Policy Paper 34), Resource Management in Asia Pacific project, and National Research Institute (Special Publication 24).

Banks, Glenn. 2003. "Landowner Equity in Papua New Guinea's Minerals Sector. Review and Policy Issues." *Natural Resources Forum* 27: 1–12.

Banks, Glenn, and Richard T. Jackson. 2011. *Xstrata Frieda River Project Draft Inward Migration Plan*. Palmerston North: Massey University, for Xstrata.

Beer, Bettina. 2008. "Buying Betel and Selling Sex Contested Boundaries, Risk Milieus, and Discourses about HIV/AIDS in the Markham Valley, Papua New Guinea." In *Making Sense of AIDS,* edited by Leslie Butt and Richard Eves, 97–115. Honolulu: University of Hawaii Press.

Beer, Bettina. 2017. "The Intensification of Rural–Urban Networks in the Markham Valley, Papua New Guinea: From Gold-Prospecting to Large-Scale Capitalist Projects." *Paideuma* 63: 137–158.

Beer, Bettina, and Tobias Schwoerer, eds. 2022. *Capital and Inequality in Rural Papua New Guinea*. Canberra: ANU Press.

Biersack, Aletta. 1996. "Making Kinship: Marriage, Warfare, and Networks among Paielas." In *Work in Progress: Essays in New Guinea Highlands Ethnography in Honour of Paula Brown Glick*, edited by Hal Levine and Anton Ploeg, 19–42. New York: Peter Lang.

Burton, John. 2014. "Agency and the «Avatar» Narrative at the Porgera Gold Mine, Papua New Guinea." *Journal de la Société des Océanistes* 138–139: 37–52.

Church, Willem. 2022. "Factional Competition, Legal Conflict and Emerging Organisational Stratification Around a Prospective Mine in Papua New Guinea." In *Capital and Inequality in Rural Papua New Guinea*, edited by Bettina Beer and Tobias Schwoerer, 63–100. Canberra: ANU Press.

Filer, Colin. 1990. "The Bougainville Rebellion, the Mining Industry and the Process of Social Disintegration in Papua New Guinea." *Canberra Anthropology* 13 (1): 1–39.

Filer, Colin, and Richard T. Jackson. 1989. *The Social and Economic Impact of a Gold Mine at Lihir: Revised and Expanded.* Port Moresby: University of Papua New Guinea.

Filer, Colin, Glenn Banks, and John Burton. 2008. "The Fragmentation of Responsibility in the Melanesian Mining Industry." In *Earth Matters: Indigenous Peoples, Corporate Social Responsibility and Resource Development,* edited by Ciaran O'Faircheallaigh and Saleem Ali, 179–196. London: Greenleaf Publishing.

Golub, Alex. 2014. *Leviathans at the Gold Mine: Creating Indigenous and Corporate Actors in Papua New Guinea*. Durham, NC: Duke University Press.

Human Rights Watch. 2010. *Gold's Costly Dividend. Human Rights Impacts of Papua New Guinea's Porgera Gold Mine*. New York: Human Rights Watch.

Jacka, Jerry. 2015. *Alchemy in the Rain Forest: Politics, Ecology, and Resilience in a New Guinea Mining Area*. Durham, NC: Duke University Press.

Jacka, Jerry K., 2019. "Resource Conflicts and the Anthropology of the Dark and the Good in Highlands Papua New Guinea." *The Australian Journal of Anthropology* 30 (1): 35–52.

Jorgensen, Dan. 2007. "Clan-Finding, Clan-Making and the Politics of Identity in a Papua New Guinea Mining Project." In *Customary Land Tenure & Registration in Australia and Papua New Guinea: Anthropological Perspectives,* edited by James F. Weiner and Katie Glaskin, 57–72. Canberra: ANU Press.

Kemp, Deanna, and John R. Owen. 2018. "The Industrial Ethic, Corporate Refusal and the Demise of the Social Function in Mining." *Sustainable Development* 26 (5): 491–500.

Koczberski, Gina, Georgina Numbasa, Emmanuel Germis, and George N. Curry. 2017. "Informal Land Markets in Papua New Guinea." In *Kastom, Property and Ideology: Land Transformations in Melanesia,* edited by Siobhan MacDonnell, Matthew Allen, and Colin Filer, 145–168. Canberra: ANU Press.

Macintyre, Martha, and Simon Foale. 2004. "Politicised Ecology: Local Responses to Mining in Papua New Guinea." *Oceania* 74 (3): 231–252.

Owen, John R., and Deanna Kemp. 2017. "Social Management Capability, Human Migration and the Global Mining Industry." *Resources Policy* 53: 259–266.

Rooney, Michelle N. 2017. "'There's Nothing Better than Land': A Migrant Group's Strategies for Accessing Informal Settlement Land in Port Moresby." In *Kastom, Property and Ideology: Land Transformations in Melanesia,* edited by Siobhan MacDonnell, Matthew Allen, and Colin Filer, 111–144. Canberra: ANU Press.

Schram, Ryan. 2014. "Notes on the Sociology of *Wantoks* in Papua New Guinea." *Anthropological Forum* 25 (1): 3–20.

Schwoerer, Tobias. 2023. "Large-Scale Land Transformations and Changing Sociality among the Wampar in Papua New Guinea." *Anthropological Forum*. Advance online publication. doi: 10.1080/00664677.2023.2247174.

Skrzypek, Emilia. 2021. *Revealing the Invisible Mine: Social Complexities of an Undeveloped Mining Project*. New York: Berghahn.

9 Mining fronts, labor mobilities, and the construction of locality in Thio, New Caledonia

Pierre-Yves Le Meur

Introduction

Mining activity disrupts and remakes places (Bebbington and Humphreys Bebbington 2018, 441). By moving mountains and people, and generating material and immaterial flows and overflows, mining destroys landscapes and de-structures societies, generating new kinds of assemblages in the form of localized, unstable, moving "minescapes." In this respect, mining can be viewed as a specific form of capitalist frontier, understood as "an imaginative project capable of molding both places and processes" (Tsing 2005, 32). However, the mining frontier process does not unfold in a social vacuum. Instead, it meets pre-existing social configurations and interacts with concomitant processes (Rasmussen and Lund 2018). The historical trajectory of New Caledonia bears the hallmark of the entanglements[1] between settler colonization, spatial and racial segregation, and mining activity. These entanglements have reshaped or produced changing governable spaces – that is, configurations of authorities, territories, identities, and resources (Watts 2004; Allen 2018). Mobility, in its various guises, has been at the heart of these processes: the imperial front of colonizing remote territories; the push toward settler colonization – either "free" or "penal" (1864–1898); the importation of labor from other French colonies and other countries such as the Dutch East Indies (now Indonesia) and Japan; land dispossession and internal displacement of Kanak people (the indigenous inhabitants of New Caledonia); the use of cattle as a weapon for encroaching on and destroying Kanak agricultural settlements; and the shifting location of mining fronts within New Caledonia, driven by the discovery of new deposits and the relative depletion of older ones.

When it comes to mining, it often feels as if everything is moving around, and nothing is stable. However, elements of stabilization occur, all the more in an island context of territorial boundedness. In this chapter, I will argue that, in the case of New Caledonia, mining activity has resulted in a double process of crystallization of specific governable spaces. At the territorial scale, the mining front has increasingly tended to function as and generate a single enclave, despite physical movements and exchanges between companies (of deposits, ore, and staff). At the local level, the politics of belonging has coalesced into a sense of locality which takes on different forms (municipality, clan network). This chapter explores this tension

DOI: 10.4324/9781003313236-12

between mobility and locality (understood in both material and representational senses), drawing mainly on the case of the locality of Thio. Located in the southern part of the east coast of the main island, Thio has been the stronghold of nickel mining and the Société Le Nickel (SLN) company for more than a century and a half.

The conceptual section that follows delineates how I will use the notions of flow, overflow, and governable space. I first position human mobility in the broader frame of material and immaterial flows and overflows that characterize extractive industries. I then introduce the concept of governable space and associated territorializing practices as a lens for analyzing the dialectic of mobility and locality in the mining context. "Locality" is a short-hand term for describing a localized material space imbued with a sense of place and belonging. The next section of the chapter presents a historical overview of the forms of mobility linked to, or induced by, mining in Thio. This overview encompasses the whole colonial era from the beginning of nickel mining in this locality in the 1870s. I will then explore how mobility has shaped residential and identity patterns in Thio (within and between ethnic groups, and along other affiliation lines) as well as local arenas and governable spaces. Focusing on recent events and conflicts, the third part of the chapter will discuss how Thio as a locality has been constructed and transformed by the interplay of mining, mobility, and place-based attachment (including customary land ownership). The chapter draws on interviews and observational data from several research projects conducted in New Caledonia in the period 2009–2019, as well as the "grey" and scientific literature. I also utilize archival material held in New Caledonia's Territorial Archives, Noumea (including SLN archives), the French Overseas Archives, in Aix-en-Provence, and Military Archives, in Vincennes.

Mining development and the mobility-locality dialectic

Flows and overflows

Mining is an activity that relies on flows, including of people, capital, machines, norms, metal, water, waste, and contaminants. The dialectic of fixity and movement that characterizes mining also lies at the core of capitalism. Drawing on Marx, David Harvey (2006, 101) argues that there is "a fundamental contradiction between fixity and movement within the theory of accumulation in space and time." This is obvious with mining development, which has been shaped by the tension between necessarily localized extraction and the movements of the mining frontier driven by geological, economic, technological, and political factors.

Large-scale mining has also generated major environmental disasters. Scholars in industrial and environmental history have labeled these adverse impacts "overflows." These include "everything that, because of the existence of a production activity and its insertion in its immediate environment, imposes its material and symbolic existence through externalities whose legitimacy is contested by the surrounding populations" (Le Roux and Letté 2013, 18). This definition underscores the political dimension of these largely uncontrolled and illegitimate flows.

The perimeter of the zone of mining impact is itself a contested notion. The perimeter is not only legal (the mining lease) and material (mining overflows) but also moral and political in nature. It includes the perimeter of corporate responsibility (and beyond), which seeks to delineate the responsibility of all the actors involved, in a context of huge power asymmetries. As Nicholas A. Bainton and Glenn Banks argue (2018, 458),

> The focus on flows and mobilities helps to sharpen our attention to the ways that mining companies sometimes attempt to 'manage' these types of 'ungovernable movements' …associated with mining and its attendant impacts, and their failure to reduce certain frictions or control immanent processes.

Governable spaces between territorializing practices and mobilities

The notion of governable space was developed by British sociologist Nikolas Rose (1999, 32) to describe "modalities in which a real and material governable world is composed, terraformed and populated." Crafting governable spaces "is a matter of defining boundaries, rendering that within them visible, assembling information about that which is included and devising techniques to mobilize the forces and entities thus revealed" (Rose 1999, 33). The concept potentially applies to any social domain or scale – a community, a factory, a family, a national economy. (See Watts 2004 and Allen 2018 for examples of how it has been applied to the extractive sector.)

This framing of governable spaces as entangled and hierarchical configurations of resources, authorities, identities, and territories is a useful analytical device for describing the New Caledonia situation. Here, the space of customary landownership takes a specific shape due to the legacy of settler colonization and still ongoing land reform. On one hand, the tribe/land reserve, which originated from the colonial policy of racial and spatial segregation, remains the structuring core of the Kanak social order at the local level and the basis for land restitution under the legal status of customary land. On the other hand, claims, conflicts, and negotiations around mining projects do not follow the same pattern. Instead, they start with a claim of recognition of autochthony or anteriority (as firstcomers) in the area concerned, regardless of the area's current legal status (generally New Caledonia's public estate for negotiations over mining projects). As I have previously noted (Le Meur 2022, 173),

> This strategy departs from the logic of claiming land for formal allocation according to the land reform policy. The claimed area is not conceptualized in the idiom and ideology of landownership … but as a space generated and actualized by the historical deployment of clan and chiefdom networks.

The mythical relation to land connecting living entities and the past underlies both processes of land restitution as customary land (in the legal sense of the 1999 Organic Law) and the negotiation of community-corporation agreements around mining projects. The space of tribe/customary land and the space of indigeneity are thus partly overlapping. However, as we will see, mining conflicts are not necessarily discursively framed in these terms and a different sense of locality has gradually emerged, from settler colonization, since the end of the *indigénat* regime[2] in 1946, and through local-level mining, land, and political conflicts. This has resulted in a trans-ethnic sense of belonging at the municipal level.

At the country level, we observe two contradictory movements: on the one hand, an opening and diversification of mining projects and companies; on the other, the increasing structuring of New Caledonia as a single mining territory. Since the 2000s, the mining sector has been structured around three large-scale mining and processing projects (see Figure 9.1). These are

i The Koniambo mine and smelter (KNS) operation located in North Province. This is majority-owned by Société Minière du Sud Pacifique (SMSP), an entity which is controlled by North Province and is the economic spearhead of Kanak pro-independence politics. Glencore, a world-leading raw material trader based in Switzerland, holds a 49 percent equity share.
ii A nickel laterite mine and refinery in the South Province, at Goro, which until late 2020 was owned by the Brazilian mining company, Vale. The mine and refinery are now 51 percent owned by a group of New Caledonian provincial authorities and other local interests, with the remaining 49 percent being held by a consortium of private companies (of which 19 percent is held by Trafigura, a Glencore rival).
iii The multi-site operations of the historical company SLN (now a subsidiary of the French mining group Eramet), which feed the Doniambo refinery on the outskirts of Noumea.

These three mining entities have been driven, to some extent, by different political, economic, and social agendas, but at the country level New Caledonia now increasingly resembles a single mining enclave (see Banks and Le Meur forthcoming on the concept of mining enclave).

One reason for this is that New Caledonia has developed strong regulations with the adoption in 2009 of the Framework for the Development of Mineral Resources (*Schéma de mise en valeur des richesses minières*) and a territorial Mining Law (*Code minier*). The Bercy agreement, signed in 1998, was the first step in institutionalizing a pro-independence vision on decolonization and mining that bore the hallmark of resource nationalism. Known since 2015 as the Nickel Doctrine, this vision is now partly shared by some contra-independence parties as well.

Other factors include the relative connectedness between mining companies (via flows of deposits, ore, and staff) and the density of infrastructural, financial, and human networks. Small and medium companies, locally known as *petits mineurs*, who are engaged in exporting ore or selling it to large-scale processing companies, contribute to the densification of the mining sector (Bouard et al. 2018). Furthermore, all potential ore bodies have been exploited at least once, and

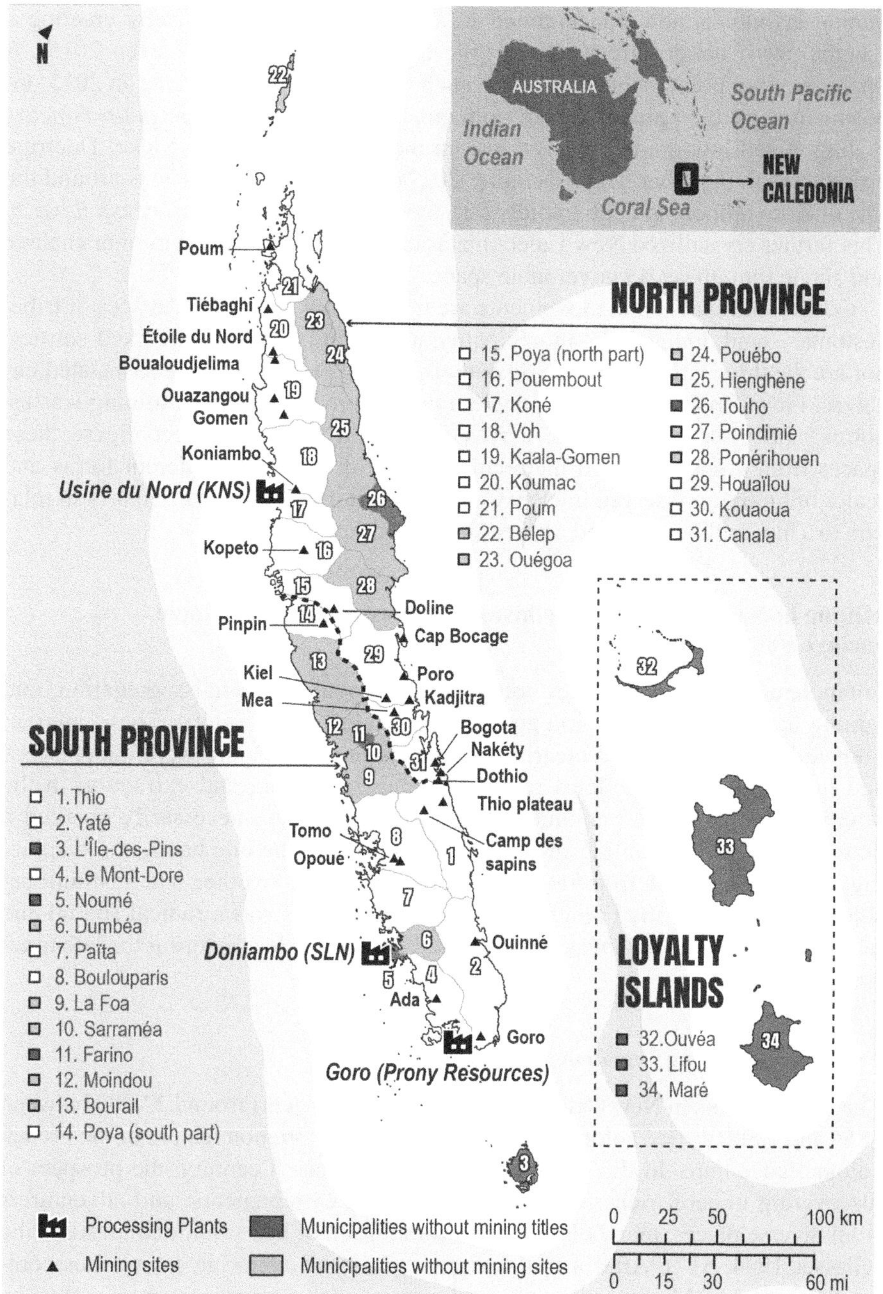

Figure 9.1 Mining geography in New Caledonia. Source: Levacher and Le Meur 2022. Map elaboration by Marilyn Ishikawa.

mining activity is now concentrated in already mined areas, thereby creating a specific post-frontier "minescape" at the New Caledonia level (Larsen 2015). A final important point is that conflicts – such as the *conflit des rouleurs* in 2015 opposing truckers of unprocessed ore from mines to coastal wharfs and *petits mineurs* willing to sell low-grade ore overseas, on the one hand, and the Nickel Doctrine promoters, on the other hand (Demmer 2017) – plus the acute tensions around the sale of Goro refinery in 2020, rapidly became national issues (*une dimension pays*). This further crystallized New Caledonia as a single (but complex) mining enclave and single (but uneasily) governable space.

Governable spaces and movements are inextricably linked. The spaces of tribe/customary land, indigeneity, municipality, and main island[3] are not fixed entities, nor are the three large-scale mining and processing projects and the associated enclaves. Flows and overflows of people, raw and processed nickel, mining wastes, norms, and discourses pervade across scales and spaces and reconfigure these spaces from a relational and hierarchical point of view. The different forms and scales of mobility observed in New Caledonia historically, and particularly in relation to Thio, will be analyzed against this backdrop.

Mining in New Caledonia: the historical construction of a single enclave

Entanglements between settler colonization, spatial and racial segregation, and mining activity have been integral to New Caledonia's history. These interactions did not result from a clearly planned strategy, and there were many twists and turns in colonial policies regarding land, migration, and extraction in the aftermath of the French conquest in 1853. Colonization necessarily entailed a double movement, enabling the influx of settlers, on the one hand, and displacing indigenous people from their traditional lands, on the other. The institutional solution favored by the French to manage this process was a radical spatial and racial segregation, which was implemented in several phases during the late nineteenth century.

Four waves of mining migration to New Caledonia

The initial settlers in New Caledonia were mainly convicts (around 22,000 between 1864 and 1898) as free colonization was a failure – a common feature of the French colonial adventure. In the second half of the nineteenth century, the prospect of discovering mineral riches attracted prospectors, entrepreneurs, and adventurers from diverse places, many of them who came from, or had sojourned in, Australia (Glasser 1904, 517). After a trial-and-error phase of exploring for various commodities (coal, gold, iron, antimony, and copper), nickel was discovered in 1864 by French engineer Jules Garnier, who was later to co-found SLN, in 1880. This soon became the main mineral resource and has been continuously exploited from 1873 until today, along with cobalt and chrome. This first movement was similar to the mining rushes that occurred in different places in the same period.

The second form of migration directly induced by mining development was the importation of indentured labor. While New Caledonia turned away from the penal strategy and failed to drive free colonization, Kanak people were increasingly locked into land reservations (and the disciplinary regime of *indigénat*). From the end of the nineteenth century, and increasingly in the interwar period, the growing labor gap that this created had to be filled through the importation of indentured labor from other colonies and countries. "The granting of the first mining concessions accounts for the considerable upsurge in the intake of New Hebrideans from 1873" (Shineberg 1999, 133). Workers later came from French Indochina (1891 onward), Japan (1892 onward), and the Dutch East Indies, especially Java (from 1896 in the agricultural sector, and from 1903 in response to a coffee crisis and new nickel boom: Devambez-Armand 1993, 212). Mining entrepreneurs played a key role in labor importation. In 1873, John Higginson, another SLN co-founder, hired the first contingent of New Hebrideans to work on the Balade copper mine in the north of New Caledonia where they worked beside convicts and skilled workers originally from English Cornwall (Thompson 2000). Twenty years later, in 1890, SLN arranged for a convoy of 599 Japanese workers to land in Thio without going through immigration service procedures. This occurred again in 1911 (Palombo 2002, 56, 67).

Given the demographic weight of immigration[4] and its importance for the functioning of New Caledonia's economy, the weakness of the legal framework in this area must be emphasized. Before the decree of 24 December 1935 that reorganized immigration in New Caledonia, the Governor pointed out in his report to the presidency of the Republic that "the immigration of exotic workers to New Caledonia has never been the object of an act of the central power" – instead, it was governed by various decrees made applicable locally for the Javanese or Indochinese immigrants by decrees of the chief of this colony. It can be argued, therefore, that the administration rationalized ex-post practices and power relations with industrialists, instead of setting in place a legal framework for economic and mining activities.

The end of the *indigénat* regime in 1946 profoundly affected the labor question in the post-World War II era, all the more as it concerned not only the indigenous Kanak but also people from French Indochina and Dutch Indonesia. The repatriation of many Indonesians and Indochinese created panic in the administration and economic world in New Caledonia. Henri Lafleur, a prominent New Caledonian politician at the time, summarized the situation well in 1947, observed, "At present, the problem of manpower is a distressing one: it is a matter of life and death for New Caledonia" (minutes of a meeting at the Ministry of Overseas France, 09/09/1947, CAOM, FM, AP, carton 509bis).

In the 1940s and 1950s, representatives of the political-administrative and economic worlds of New Caledonia struggled to find external sources of manpower from places like Indonesia, Fiji, China, Japan, Morocco, and Italy (various files, CAOM, FM, AP, cartons 509 and 509bis). Various reports prepared for the centennial of colonization in 1953 emphasized the scarcity of labor, notwithstanding progress toward mechanizing the mining industry and an increase in the number of people leaving cattle stations to work in mines or in the capital, Nouméa. In response to a reduction in Indonesian and Vietnamese immigration, an August 1953 report of the Caisse centrale de la France d'Outre-mer (CAOM, FM, AP, carton 509) proposed importing French and European workers (northern Italians in particular) and incorporating more New Caledonians into production (although the same report mentioned the "natural indolence" and "mediocre attendance" of the "Indigènes" who "prefer tribal life on the reserves to mine work"). Another document prepared for the visit of Jacquinot, Minister of Overseas France, entitled "The Human Problem," was a vibrant plea for "white immigration" as a response to the closing of the Asian labor markets.

This situation triggered a third wave of migration. In the 1950s and 1960s, people mainly came from other French ex-colonies (Morocco, Algeria) and remnants of the imperial space (Wallis-and-Futuna, French Polynesia). This process culminated with the 1967–1972 nickel boom, which saw a renewal of settler colonization aimed at demographically drowning-out any Kanak claim for independence.[5] The 1967–1972 nickel boom and the associated flow of newcomers (many of whom remained in New Caledonia) can be interpreted as the delayed achievement of nineteenth-century settler colonial policy.[6]

The end of the *indigénat* regime offered a new clientele to the unions, who viewed with disfavor SLN's plan to import Italian labor (Israël 2007, 60). The law of wage parity between "Europeans" and other ethnic groups was finally passed in 1957. As Benoît Trépied (2010, 259) observes in his fine-grained historical ethnography of the municipality of Kone in the north of New Caledonia

> [T]he process of mechanization of production, as well as the intersecting social mobility of Europeans in the process of proletarianization and Asians leaving their former status of indentured workers, led to a gradual standardization of the condition of worker-miner – which had been particularly contrasted until then because of the statutory and racial cleavages of the colonial era – in post-war New Caledonia.

A fourth wave of "temporarily imported" Asian labor took place in the 2000–2010s, with the construction of two large nickel processing plants in Goro, in the South, and Vavouto (the Koniambo Project), in the North, starting in 2005 and 2007, respectively. A skilled workforce was temporarily hired on the two construction sites as part of contracts with Asian companies in charge of building and assembling the plants (Merle 2012, 4–6). The two construction sites each hosted around 7,000 workers at their highest, most of whom were from Asia (China, South Korea, Thailand, and the Philippines).

By this time, the Organic Law of 19 March 1999 had transferred full jurisdiction over labor law to New Caledonia. The Law also provided for the drafting of a law (*loi du pays*) protecting local employment (employment of New Caledonian citizens) conceived as a key element of the negotiated decolonization process underlying the 1998 Noumea Accord. The first edition of the New Caledonia labor code was published in 2008, and the law on local employment was finally issued in

2010. In 2002, a law was passed aimed at controlling the workforce of international contractors in the context of a new nickel boom that had reignited fears of state-sponsored immigration.

Internal mining fronts and mobilities

The migratory waves triggered by mining activity have shaped New Caledonia's political economy and settler colonization. Mine-associated mobilities have also been influenced by the mineral geography of New Caledonia, namely the location, accessibility, and content of ore deposits (see Figure 9.1). The evolution of the sector has been driven by external factors (international price and demand) and internal dynamics around the discovery of ore deposits and the capacity to process nickel in-country.

The opening and closing of mines defined patterns of residence and mobility. Few mining sites were continuously exploited following their discovery, the Plateau mine in Thio being a rare exception. Thus, one observes recurring moves of people from one site to another. For instance, when Koniambo was closed in 1947, many families migrated to Thio (Merle 1995; Trépied 2010). These movements also produced mining villages around mining enclaves such as Tiebaghi (Jouve 2009) in the post-war era or later Kouaoua, the last mining town – or rather SLN town – in New Caledonia. This point is explored further in the next section devoted to Thio.

Over time, the movements of internal mining fronts have produced or strengthened localities. Mining companies have also taken on state functions, as illustrated by the case of SLN. The company acted to a certain extent as a substitute for the state, adopting a paternalistic character (Reid 1985; Noiriel 1988), in a context marked by the financial and administrative weakness, and the repressive brutality, of the colonial state. SLN provided a large part of the infrastructure: the first hospital in the 1880s, schools on mining compounds, a bakery, a butcher's shop, sport associations, and later a housing estate and electricity (Le Meur 2013). This pattern can be observed in other mining towns in New Caledonia where mining companies also play a key role in the structuring of local governable spaces. They have indirectly contributed to the local anchoring of a state apparatus, functionally, as said, but also from a historical vantage point. For instance, SLN assets such as electric networks were communalized in the 1960 and 1970s alongside the emergence of the municipal government, and later with the provincialization promulgated by the 1988 Matignon-Oudinot political agreements.

Mine-related mobility must also be understood within broader dynamics linking activities at different scales, from the household to the level of New Caledonia as a whole. Labor mobility includes how individuals and households manage a plurality of activities: (i) *synchronically* – combining different economic activities in the same period; and (ii) *diachronically* – switching from one sector to another one, and thus another place, according to economic opportunities and social expectations. Livelihoods in New Caledonia combine market and non-market elements (including community activities and customary exchanges) and agricultural (including hunting and fishing) and non-agricultural activities, of which mining, through wage labor and/or entrepreneurship (sub-contracting), is one component (Gaillard and Sourisseau 2009; Apithy et al. 2016). Movements between agriculture and mining in response to the booms and downturns of the nickel sector (Gaillard 2010) have not entailed the end of non-market practices. To the contrary, the active maintenance of these practices, and the social bonds within which they are embedded, is the expression of an anti-risk strategy. Similarly, movements toward cities (mainly Noumea and its suburbs, and more recently Voh-Koné-Pouembout (VKP) with the KNS project) have not weakened social and kinship ties with tribal areas. Greater mobility has also led to increased circulation of money and products between urban and tribal areas (Apithy et al. 2016, 324). What we observe, therefore, is the development of strategies of multiple residencies or "binding mobilities" (Pestaña, Hoffer, and Pantz 2016, 368–372), which maintain or generate ties and networks, of which the mining sector is one component and site.

All these flows, movements, and mobilities, combined with the political and administrative apparatus generated by the political agreements of Matignon-Oudinot (signed in 1988) and Noumea (signed in 1998), have contributed to the historical production of New Caledonia as a specific governable space at the territory level. This lends support to the hypothesis of New Caledonia as a single enclave. We now move down to the local level to explore in more detail how mining contributes to, and is structured by, the mobility-locality dialectic.

Thio: mobility, overflow, and the gradual construction of locality

The geography of Thio, located in the southern part of the main island's east coast, displayed features that facilitated mining development, such as very rich nickel deposits with an easy link to the sea via two valleys and a coastal plain, although there were also adverse elements to contend with, such as shallow coastal waters, coral reefs, and exposure to prevailing winds (Bencivengo 2010, 710). As indicated, Thio became the stronghold of SLN, a nickel company that long held a hegemonic position in the New Caledonian mining landscape.

The history of Thio embodies at different times all the forms of mining and mining-induced mobility to be found in New Caledonia:

Inflows of prospectors and entrepreneurs.

Foreign labor force recruitment from the end of the nineteenth century.

Displacement of Kanak people due to extraction and infrastructure construction and neglect – the two sides of the same coin of pervasive infrastructural violence (Rodgers and O'Neill 2012).

Internal residential movements related to wage-labor opportunities at the mine and access to SLN housing schemes.

Medium-range workforce movements in and out of Thio according to the cycles of mine closure and opening at the country level (especially during and after the nickel boom of 1967–1972).

People choosing to exit the mining sector for political-environmental reasons in the 1990s or trying to get control over time by resorting to the moratorium instrument in the 2010s.

Paradoxically, these complex mobility forms and their interplay with mining development have slowly created a sense of locality at the municipal level, combining issues of land control, belonging, and local citizenship (Jacob and Le Meur 2012). This sense of locality is strongly but ambivalently linked to the mining issue, its infrastructural anchoring, and specifically to SLN, which positions and claims itself to be "Caledonian" and is locally challenged by people inviting "it" to be a good "local citizen." In recent years, in conflict contexts, the rhetoric of claims has moved from a Kanak/customary discourse (visible in the conflicts of the 1990s; see Le Meur 2017) toward the assertion of a municipal and trans-ethnic citizenship, as well as from a claim to land ownership (via the land reform process) to a more explicit assertion of sovereignty (Le Meur 2022).

Over time, mining has strongly contributed to the structuring of governable spaces in Thio, and SLN has become one of the main centers of power within local arenas. The three case studies briefly presented in this section together contribute to sketching the complex political landscape of Thio, which comprises plural, partially overlapping governable spaces and is structured by various interactions between locality and mobility. Each case approaches the governability of spaces from a specific vantage point: a mining enclave (the Plateau mine), a tribal locality (Ouroue), and a municipality (Thio).[7] One important question for the analysis is the degree of autonomy (or its converse, heteronomy) of these places in their relation to broader scales of government and especially to New Caledonia understood as a single mining enclave.[8]

Plateau mine: from settler colonization to Kanak emancipation

The Plateau mine, which is actually a block of several mining leases, came to be a key source of nickel – accounting for around 70 percent of all nickel ores extracted in the Thio mining center and around 20 percent of all the ore mined on the main island from the 1870s until the end of the 1980s (Bencivengo 2010, 710). The Plateau is a pivotal component of a wider "minescape" that progressively developed around SLN. It included the creation of Thio village by the colonial administration in 1879, one year before SLN's birth. The company soon became the main driver of the development of the village, owning 30 percent of the built land plots in the 1880s and 50 percent by the 1930s. The company also established its head office there in the 1920s. In 1887–1888 SLN implemented a mining camp for convicts at the heart of the Plateau for the sake of disciplinary control, while other worker camps were located at the foot of the mines (Bencivengo 2010, 895–901). These mining camps formed part of a dense web of infrastructure for transporting nickel from the mine to the boat and to the smelter that had been built in Thio Mission in 1912. The Plateau epitomized the labor policy carried out by the colonial administration and the mining sector, sometimes jointly and sometimes separately.

It brought together the different workforce layers, comprising mostly imported workers – from New Hebrides, Japan, French Indochina, Dutch Indonesia – and convicts and freed convicts, all of whom worked at the mine.

Mining development at the Plateau site generated environmental damage and overflows that impacted the daily life of neighboring tribes, especially Saint-Philippo 1 located in the floodplain at the foot of the mine. Overflows were of different kinds, including not only rock and water but also humans, especially the penal workforce with whom frictions occurred but who also developed social ties that were frowned upon by the colonial administration (see Muckle 2011, 148–154).

Displacement became a key feature of the residential patterns of tribes such as Saint-Philippo 1 due to mining overflows. In the 1950–1970s, the environmental impact of mining increased dramatically due to the rapid mechanization of extraction and the lack of environmental regulations. Saint-Philippo 1 inhabitants increasingly suffered from flooding (resulting from cyclones, like Allison in 1975) and started to move away from the Thio River by the 1950s.[9] Some of them left the tribe, which was relocated three times on both sides of the river in a time span of 20 years. As long as these resettlements occurred within the limits of the tribal land reserve, they went unnoticed by the SLN (being regarded as "externalities"), although residents eventually received petty financial compensation for rehousing in non-flood areas from the company in the 1980s, under the so-called "*pieds mouillés*" ("wet feet") scheme.[10]

Parallel to these internal displacements, Kanak came from other parts of the territory to work on the Plateau before and mostly after World War 2. A large number came from the Loyalty Islands, in particular Lifou and Ouvéa.[11] These workers inhabited an encampment beside the Plateau mine up to the 1970s and later lived in SLN housing estates near the wharf at Pawani. A new wave of Kanak workers arrived during the 1967–1972 nickel boom to work for mining companies (not just SLN but also *petits mineurs* that rapidly grew and multiplied during the boom) or subcontractors, many of which were companies of French origin.[12] Some of these workers subsequently settled in Thio permanently, married, and eventually became involved in local associations or pro-independence movements. The *Association des Kanak de la Grande Terre et des Îles* (Associations of Kanak from the Main Island and the Loyalty Islands) emerged from these migratory movements. This "ethnonational" association, initially led by pro-independence militants, later evolved into a neighborhood association, the *Association du Botaméré*, which took its name from the sacred sugarloaf-shaped mountain overhanging the Thio Mission estate. This reflects tensions between Kanaks native to Thio and those who were not that were generated by the post-nickel boom economic downturn.

Later, in the 1990s, the Plateau became one of the most symbolic sites for the Kanak who had pushed the mining issue to the forefront of the pro-independence struggle (Bouard et al. 2016; Le Meur and Levacher 2022). During a 1996 conflict that proved to be pivotal to relations between the Kanak community of Thio and the SLN, the entry-gate to the Plateau mine was chosen as the place for negotiating a broad agreement with the SLN and the political authorities (Le Meur 2017). This was a material expression of the governable space structured by mining activity in the locality of Thio. In short, the trajectory of the Plateau illustrates the relations and ramifications – mobilities, flows, and overflows – of organizing a mining enclave that has also become over time a place loaded with symbolic and political value.

Ouroue tribe, a history of mine-induced displacements: "Imagine if we all moved, where would we go?"[13]

Ouroue tribe is located on the coast, west of Thio village. Its history embodies the different forms of localized mobilities associated with movements of the mining front. The alluvial plain of the Dothio River quickly attracted settlers, but relationships with neighboring indigenous people rapidly deteriorated and they left the place. The administration was reluctant to dispossess natives who were their allies during the 1878 rebellion. Governor Olry was opposed to a first project of reserve delimitation, only granting one to two ha per person to the natives. The first delimitations of land reservations occurred in 1880 in Thio (Dauphiné 1989, 139). The decree of 15 March 1880 defined six land reserves in Thio, understood as a response to the "needs of the natives." Ouroue tribe (the administrative notion of "tribe" was created in 1867) was granted 116.10 ha. However, in 1887, SLN obtained a permit to establish a tramway bringing nickel from the Pauline mine to the coast. This became operational in 1888. In the same year, 93 convicts settled near Ouroue to work at the mine. In 1889, SLN built a nickel processing plant at Ouroue. The negotiation with the Ouroue chief Pita resulted in the expropriation of more than 116 ha (decree of 24 May 1889). The refinery closed two years later because of a decline in nickel prices and the bad location of the building, which had been constructed on unstable sandy soils. In the meantime, Ouroue inhabitants had fled back to their previous location in the hills surrounding the upper Dothio River. A settlement center was planned in 1895–1896 in the Dothio Valley. This failed completely, and a new indigenous reserve was created in 1899–1900 in Dothio-Ouroue, setting aside 118 ha for 34 inhabitants. The reserve lost 9.25 ha in 1913, which was attributed to the incursions of a white settler, but increased its size in 1941, 1960, and 1975 (respectively by 3.40 ha, 25.00 ha, and 5.15 ha; ORSTOM 1980).

At the same time, the Ouroue environment suffered from adverse mining impacts: tailings were dumped from the Plateau and Wellington mines into Wellington Creek (*xwâ Nêmûrû*), where people had previously fished. The opening of the Petrel mine in the mid-1960s generated further pollution and created problems of access to water, which led to blockades being organized by the Ouroue population. Eventually, SLN stopped exploiting Petrel and compensated local residents by providing financial assistance to build houses in a new location, called La Plaine (the 5.15 ha mentioned above). However, there were recurrent demonstrations against the damage caused to Wellington Creek, which looks like a river of rocks nowadays (see Figure 9.2). In the meantime, thanks to clanship ties with the Tura (the chieftaincy clan in the lower valley of Thio), two of the three clans had moved to Nembuya, located eastward between the Thio and Nepu Rivers, closer to the sea. This area would be increasingly affected by mine impacts and wastes.

Figure 9.2 Wellington Creek. Photo by author (October 2013).

Mine-induced population movements – that is, overflows – strongly affected the trajectory of Ouroue tribe, which had already been impacted by settler colonization from the late nineteenth century onward. Individual responses were influenced by clanship affiliations and histories. There were also attempts to leave the current (and small) residential area of La Plaine to go back to the "old tribe" (actually delimited during the colonial time but where the cemetery is located) or even to the Dothio area further west in the hills. SLN was seen as the key negotiation partner, and a local agreement was signed in 2007 between Ouroue tribe and the company. In fact, having obtained an agreement similar to that of 1996 mentioned above (Le

Figure 9.3 Discussion between Ouroue and Thio leaders and representatives of the SLN Thio Mines. Photo by author (September 2014).

Meur 2017), Ouroue leaders had to "increase the pressure again" on SLN for the agreement to be implemented, to the extent that new negotiations had to be held and a new text put on the table in 2007.

Overflows came to the fore again in the 2010s, when SLN started discussions with Ouroue authorities to launch a drilling campaign in the Wellington Creek tailings to assess the potential for nickel re-exploitation (see Figure 9.3). The negotiation – tense and conflictive at some points – revolved around the land-mine nexus: the tailings are partly located on customary land and extracting nickel from this area – which required re-designating mining overflows or tailings as a mine – raised questions about benefit sharing.[14] The matter is still pending.

Chavaa xûâ association: the production of locality

In 2013, two exceptionally intense rainfall events occurred in Thio within a few months of each other – Cyclone Freda in January and torrential rains in June. The floods and landslides caused material damage in the localities located at the bottom of the slopes and downstream from the rivers. The population of Thio mobilized very quickly by blocking access to the three mining sites in the commune (Plateau, Camp des sapins, and Dothio). A list of demands was drawn up and a collective was created, explicitly interethnic in composition and including SLN employees.

A memorandum of understanding was signed a few weeks later by the collective, the town hall, the South Province, the New Caledonian government, and SLN. The collective subsequently formed an association whose name, Chavaa xûâ, means "to take care of one's house/community" in the Xârâcùù language. In addition, a simplified joint stock company (SAS) was created to organize the interface between SLN and local contractors for the maintenance and environmental rehabilitation projects discussed under the agreement.

Immediately, a question arose regarding monitoring the necessary restoration and environmental maintenance work. The meetings organized in the affected areas involved the signatories to the agreement and the Nickel Fund, a New Caledonian public institution, one of whose missions is the rehabilitation of sites degraded by mining activity. The issue of follow-up, it became clear, could be resolved according to two non-exclusive narrative options: (1) through technical follow-up of the work carried out; or (2) through an inclusive political logic of local citizenship. In other words, Chavaa xûâ's strategy combined collective action on the ground, aimed at putting pressure on ecological restoration actors, in particular SLN, and a political and moral discourse, urging SLN to be consistent with its communication campaigns featuring SLN as a "Caledonian company." SLN, in this way, was reminded of its duties as a "local citizen," of its moral obligation to take into account the degradation of the environment beyond its formal mining holdings (the legal mining leases) and even, implicitly, its own direct impact. Corporate social responsibility was re-localized and oriented "inward," by and toward the Thio audience: SLN as a good local citizen rather as a responsible global corporation.

The chain of events triggered by the 2013 storms intersected with other recent developments, in particular those generated by the implementation of the 2009 New Caledonian Mining Code regulations – which involved moving from a declaratory to an authorization regime – and the desire of several mining companies to launch new prospecting campaigns on the Forgotten Coast (see footnote vii). The intersection of recent events (rains, mining code, prospecting campaigns) was interwoven with older chains of events and memories, part of the long colonial period and organized around the policy of spatial and racial segregation linked to settlement colonization, as well as the parastatal functions assumed by SLN. The discourse of Chavaa xûâ was explicitly inclusive and strove to overcome the fault lines that ran through the populations mobilized in Thio, between Kanak and non-Kanak people, between SLN and its own employees mobilized in the interethnic collective, between the communal-level claimed by the collective and the micro-local level of the tribe (for accessing public work contracts), and between generations (the Kanak "youth" appearing more radical than their elders or more prone to flee the mining sphere for environmental reasons).

The membership of the collective who signed the memorandum of understanding on 11 July 2013 gathered all the Kanak tribes and customary districts of Thio, the SLN villages (Pétroglyphes, Pawani, Bota Méré), Thio village, and the localities of Caldoche (Dallas) and Wallisian and Futunian (Bourgade). The gradual coalescence of Thio as a locality and governable space also resulted from the active political engagements of local leaders who could have sat on the opposite sides of the frontline in the 1980s civil conflict euphemistically called "the events." In September 2014, during a post-agreement visit on site to assess the environmental work being launched, I could see a former Caldoche (descendent of white settlers, partly convicts), an anti-independence activist (who had left the place in the late 1980s to come back in the 2010s), and a pro-independence, indigenous activist and trade-unionist walking hand-in-hand and claiming they were all "sons of Thio." The production of a multi-ethnic locality thus resulted from the interplay of mining overflows and multiple layers of mobilities – from colonial indentured labor and settler colonization; to the arrival of Wallisian and Futunian, mainly from the 1950s onward, but beginning earlier (Tāvaka Committee 2009, 70–75); to the state-induced French immigration during the 1967–1972 nickel boom; to the internal displacement of Kanak tribes due to the environmental damage caused by mining, which contributed to the out- and in-migration of Kanak from or to various places.

Conclusion

The inescapable observation that mining "disrupts and remakes places" (Bebbington and Humphreys Bebbington 2018, 441) draws attention to the massive overflows and indirect impacts on neighboring communities that this activity generates. As I have sought to demonstrate here, a full analysis of the mining-mobility nexus needs to consider all the forms of human and material (nonhuman) mobilities, flows, and overflows triggered by mining activities.

In the case of New Caledonia, new and reconfigured localities have progressively emerged from this destructive/productive interplay between mining fronts, labor mobility, land ownership, and place-making. These localities have crystallized around a sense of belonging and local citizenship and infrastructural movements, including the material dimension of mining and associated mobilities. Mining has thereby created new spaces of governability, as well as of conflict and ungovernability.

It is also important to note that the New Caledonian mining sector has not been the only actor in this process. The process of creating new localities was (and is) also part of a specific trajectory of settler colonization. At the country scale, mobility has been shaped by the complex history of mining fronts moving to the rhythm of the opening, closing, and re-opening of mining sites, changes which were in turn driven by technological, economic, and political forces. SLN was the dominant company for over a century, and it contributed to more than company-towns such as Kouaoua. The company was also instrumental in the genesis and maintenance of a "company-territory," as long as its hegemony lasted. The last two decades have seen the arrival of two large-scale projects in the North and South Province, the issuing of a mining code and a Framework for the Development of Mineral Resources, and the emergence and diffusion of a strategy aimed at controlling the added value generated by nickel processes. All this, combined with the density of infrastructural and staff networks and flows of deposits and ores between companies and sites, has reinforced the construction of New Caledonia as a single "enclave" (see also Boudjema and Bouard, Chapter 10 in this volume).

A final observation is that this "enclave" is far from homogeneous, hence the need to also explore the local level. What the case studies from Thio show is how local governable spaces structured around mining have been materially and politically reshaped by the overflows of mining and the attempts to contain them and to secure compensation for those who have been impacted. These processes have crystallized in localized belongings and evolved through variable connections and linkages with supra-local actors and institutions. These heteronomous forms of territory, locality, and citizenship contribute from below to the construction of a country awaiting decolonization.

Notes

1 I follow Tim Ingold who stresses intricacy and indeterminacy in his definition of entanglements: "Not a network of connections but a meshwork of interwoven lines of growth and movement" (Ingold 2010, 3).
2 An expression of colonial despotism, the *indigénat* regime, created in 1887, was a list of repressive measures applied to colonial subjects, namely the indigenous Kanak people and imported indentured workers. It was intended as provisional but was regularly reviewed and strengthened until its abolition in 1946 (Merle and Muckle 2019).
3 *Grande Terre* in French: the name of the main island where nickel is mined.
4 The Asian workforce represented 15–25 percent of the New Caledonia population in the 1920s–1930s, with a peak of 14,535 in 1929 (Bencivengo 1999, 64).
5 In the long term, the indigenous nationalist claim will only be avoided if the communities not originating from the Pacific represent a majority demographic mass" (letter from P. Messmer to Jean-François Deniau, Secretary of State for the DOM-TOM, 19/07/1972).
6 This disarticulated trajectory can be compared to the four-phase model of settler colonization elaborated by New Zealand historian James Belich (2009) for the "Anglo-World."
7 A fourth case, that could not be included for reasons of space, is the "Forgotten Coast" which runs from the south of Thio to the neighboring commune of Yaté. This area, "forgotten" by public infrastructures (there is no road between the two commune) was exploited until the early 1980s and coveted again in the 2010s by several mining companies. See Levacher and Le Meur (2022) for a more detailed presentation of this historical "minescape."
8 I follow here Penelope Anthias' analysis of the genesis of a heteronomous hydrocarbon citizenship in the Bolivian Chaco. Anthias (2018, 205–206) shows how native community lands have been "resituated and resignified in the context of multi-scalar conflicts over the governance of gas," reflecting "indigenous leaders' efforts to exercise agency in constrained circumstances and to position themselves amidst a broader set of political and territorial struggles."
9 Various interviews in Saint-Philippo 1 in 2009–2010.
10 Protocol agreement between Saint-Phillipo 1 Tribe and the SLN, 2 January 1984.
11 *Rapport Gayet pour le Service des affaires indigènes, main d'œuvre océanienne* 1929, Colonial Archives in Aix-en-Provence, *Fonds ministériels, Affaires politiques 746, Missions Revel 1922 et Coste 1929.*
12 It is worth noting that in the same period, Kanak (mostly men) from Thio went to Kouaoua, located on the East coast, north of Thio, to work on other (SLN) mining sites like Kiel, thereby contributing to the movement of the internal mining fronts mentioned above (various interviews with Kouare tribe elders, Thio, 2008).
13 Ouroué tribe's chief and his wife, Ouroué, interview 30 September 2013.
14 Various interviews with Ouroue leaders and SLN representatives, 2013–2014 (NERVAL Project).

References

Allen, Matthew. 2018. *Resource Extraction and Contentious States. Mining and the Politics of Scale in the Pacific Islands*. London: Palgrave Macmillan.

Anthias, Penelope. 2018. *Limits to Decolonization. Indigeneity, Territory, and Hydrocarbon Politics in the Bolivian Chaco*. Ithaca, NY and London: Cornell University Press.

Apithy, Leïla, Séverine Bouard, Samuel Gorohouna, Stéphane Guyard, and Jean-Michel Sourisseau. 2016. "Dynamiques économiques et sociales des tribus et de la ruralité: fragilités et facultés d'adaptation." In *La Nouvelle-Calédonie face à son destin: Quel bilan à la veille de la consultation sur la pleine souveraineté?* edited by Séverine Bouard, Jean-Michel Sourisseau, Vincent Géronimi, Séverine Blaise, and Laïsa Ro'i, 315–352. Paris: IAC-Karthala-GEMDEV.

Bainton, Nicholas, and Glenn Banks. 2018. "Land and Access: A Framework for Analysing Mining, Migration and Development in Melanesia." *Sustainable Development* 26 (5): 450–460.

Banks, Glenn, and Pierre-Yves Le Meur, forthcoming. "Closure and connection: A Southwest Pacific reappraisal of the mining enclave." In *Mining Encounters and Indigenous Sustainable Livelihoods: Perspectives from the Americas, Fennoscandia and Oceania*. Abingdon: Routledge.

Bebbington, Anthony, and Denise Humphreys Bebbington. 2018. "Mining, Movements and Sustainable Development: Concepts for a Framework." *Sustainable Development* 26 (5): 441–449. doi: 10.1002/sd.1888.

Belich, James. 2009. *Replenishing the Earth: The Settler Revolution and the Rise of the Anglo-World, 1783–1939*. Oxford: Oxford University Press.

Bencivengo, Yann, ed. 1999. *101 mots pour comprendre: La mine en Nouvelle-Calédonie*. Nouméa: Editions Ile de Lumière.

Bencivengo, Yann. 2010. *La Société Le Nickel. Une entreprise au cœur de la naissance de l'industrie du nickel (1880–1914)*. PhD Diss., Paris: Université Paris 1 Panthéon-Sorbonne (3 volumes).

Bouard, Séverine, Jean-Michel Sourisseau, Vincent Géronimi, Séverine Blaise, and Laïsa Ro'i, eds. 2016. *La Nouvelle-Calédonie face à son destin: Quel bilan à la veille de la consultation sur la pleine souveraineté?* Paris-Pouembout: IAC-Khartala-GEMDEV.

Bouard, Séverine, Claire Levacher, Yann Bencivengo, Louise Decottigny, Christine Demmer, Pierre-Yves Le Meur, Séverine Blaise, John Burton, Franck Enjuanes, and Sonia Grochain. 2018. *Petites et moyennes entreprises minières en Nouvelle-Calédonie*. Final unpublished research report, Noumea, New Caledonia: CNRT Nickel et son environnement.

Dauphiné, Joël. 1989. *Les spoliations foncières en Nouvelle-Calédonie (1853–1913)*. Paris: L'Harmattan.

Demmer, Christine. 2017. "L'export du nickel au cœur du débat politique néo-calédonien." *Mouvements* 91: 130-140. doi: 10.3917/mouv.091.0130.

Devambez-Armand, Véronique, 1993, "Les recrutements: Chronologie de la main d'œuvre immigrée sous contrat en Nouvelle-Calédonie." In *Le peuplement du Pacifique et de la Nouvelle-Calédonie au XIXx siècle: Condamnés, colons, convicts, coolies, Chân Dang*, edited by Paul De Deckker, 208–217. Paris: L'Harmattan-Université Française du Pacifique.

Gaillard, Catherine. 2010. *Développement minier et ruralité en Nouvelle-Calédonie*, Unpublished research report, Pouembout, Nouvelle-Calédonie, IAC (Neo-Caledonia's Agronomic Institute).

Gaillard, Catherine, and Jean-Michel Sourisseau. 2009. "Système de culture, système d'activité(s) et *rural livelihood*: Enseignements issus d'une étude sur l'agriculture kanak (Nouvelle-Calédonie)." *Journal de la société des Océanistes* 129 (2): 279–294.

Glasser, Emile. 1904. *Rapport à Monsieur le Ministre des colonies sur les richesses minérales de la Nouvelle-Calédonie.* Paris: Dunod.

Harvey, David. 2006. *Spaces of Global Capitalism: Towards a Theory of Uneven Geographical Development.* London and New York: Verso.

Ingold, Tim. 2010. "Bringing Things to Life: Creative Entanglements in a World of Materials." ESRC National Centre for Research Methods, NCRM Working Paper Series 05/10.

Israël, Henri. 2007. *Une histoire du mouvement syndical en Nouvelle-Calédonie.* Nouméa: Editions Île de lumière.

Jacob, Jean-Pierre, and Pierre-Yves Le Meur. 2012. "Three Views of a Secret: Land Rights and the Constitution of Local Citizenship." *Africa Spectrum* 47 (1): 89–96.

Jouve, Epone. 2009. *Tiébaghi 1945–1961. Mémoires d'un village minier.* Noumea: Ed. Association pour la Sauvegarde du Patrimoine Minier et Historique du Nord Calédonien (ASPMHNC).

Larsen, Peter Bille. 2015. *Post-Frontier Resource Governance. Indigenous Rights, Extraction and Conservation in the Peruvian Amazon.* Houndmills: Palgrave Macmillan.

Le Meur, Pierre-Yves. 2013. "Locality, Mobility and Governmentality in Colonial/Postcolonial New Caledonia: The Case of the Kouaré Tribe (*xuâ Xârâgwii*), Thio (*Cöö*)." *Oceania* 83: 130–146. doi: 10.1002/ocea.5009.

Le Meur, Pierre-Yves. 2017. "Conflict and Agreement: The Politics of Nickel in Thio, New Caledonia." In *Large-Scale Mines and Local-Level Politics: Between New Caledonia and Papua New Guinea*, edited by Colin Filer and Pierre-Yves Le Meur, 157–182. Canberra: ANU Press.

Le Meur, Pierre-Yves. 2022. "Proof and Test. The Construction of Customary Land in New Caledonia." In *Spaces of Custom and Law*, edited by Edoardo Frezet, Marc Goetzmann, and Luke Mason, 160–178. Abington: Routledge.

Le Meur, Pierre-Yves, and Claire Levacher. 2022. "Mining and Competing Sovereignties in New Caledonia." *Oceania* 92 (1): 74–92.

Le Roux, Thomas, and Michel Letté, eds. 2013. *Débordements industriels: Environnement, territoire et conflit (XVIII*e*-XXI*e* siècle).* Rennes : Presses Universitaires Rennes.

Levacher, Claire, and Pierre-Yves Le Meur, 2022, "The Compensation Arenas in South New Caledonia. Minescape, Governmentality and Politics." *Extractive Industries and Society* 11: 100999. doi: 10.1016/j.exis.2021.100999.

Merle, Isabelle. 1995. *Expériences coloniales. La Nouvelle-Calédonie. 1853–1920.* Paris: Belin.

Merle, Isabelle. 2012. *Autour de la condition ouvrière en Nouvelle-Calédonie. Le recrutement des travailleurs asiatiques entre les deux guerres et aujourd'hui sur les chantiers de mine.* Working Paper n°12, Programme 'Gouvernance minière', Nouméa : CNRT Nickel et son environnement.

Merle, Isabelle, and Adrian Muckle. 2019. *L'indigénat. Genèses dans l'empire français. Pratiques en Nouvelle-Calédonie.* Paris: CNRS éditions.

Muckle, Adrian. 2011. "'Natives', 'Immigrants' and '*Libérés*': The Colonial Regulation of Mobility in New Caledonia." *Law Text Culture* 15: 135–161.

Noiriel, Gérard. 1988. "Du patronage au paternalisme. La restructuration des formes de domination de la main d'œuvre ouvrière dans l'industrie métallurgique française." *Le Mouvement social* 144: 17–35.

ORSTOM. 1980. *Evolution de l'espace foncier mélanésien.* 6 volumes. Paris: ORSTOM-MEDETOM.

Palombo, Philippe. 2002. *La présence japonaise en Nouvelle-Calédonie (1890–1960). Les relations économiques entre le Japon et la Nouvelle-Calédonie à travers l'immigration et l'industrie minière*. Lille: ANRT.

Pestaña, Gilles, Oliver Hoffer, and Pierre-Christophe Pantz. 2016, "Mobilités, dynamiques territoriales et urbaines." In *La Nouvelle-Calédonie face à son destin: Quel bilan à la veille de la consultation sur la pleine souveraineté?* edited by Bouard, Séverine, Jean-Michel Sourisseau, Vincent Géronimi, Séverine Blaise, and Laïsa Ro'i, 353–390. Paris: IAC-Karthala-GEMDEV.

Rasmussen, Mattias Borg, and Christian Lund. 2018. "Reconfiguring Frontier Spaces: The Territorialization of Resource Control." *World Development* 101: 388–399.

Reid, Donald. 1985. "Industrial Paternalism: Discourse and Practice in Nineteenth-Century French Mining and Metallurgy." *Comparative Studies in Society and History* 27 (4): 579–607.

Rodgers, Dennis, and Bruce O'Neill. 2012. "Infrastructural Violence: Introduction to the Special Issue." *Ethnography* 13 (4): 401–412.

Rose, Nikolas. 1999. *Powers of Freedom*. Cambridge: Cambridge University Press.

Shineberg, Dorothy. 1999. *The People Trade. Pacific Island Laborers and New Caledonia, 1865–1930*. Honolulu: University of Hawai'i Press.

Tāvaka Committee. 2009. *Tāvaka lanu'imoana – Mémoires de voyage*. Nouméa: ADCK-Tavaka.

Thompson, Anne-Gabrielle. 2000. *John Higginson. Un spéculateur-aventurier à l'assaut du Pacifique: Nouvelle-Calédonie, Nouvelles-Hébrides*. Paris: L'Harmattan.

Trépied, Benoît. 2010. *Une mairie dans la France coloniale, Koné, Nouvelle-Calédonie*. Paris: Karthala-CERI.

Tsing, Anna Lowenhaupt. 2005. *Friction: An Ethnography of Global Connection*. Princeton, NJ and Oxford: Princeton University Press.

Watts, Michael, 2004. "Resource Curse? Governmentality, Oil and Power in the Niger Delta, Nigeria." *Geopolitics* 9 (1): 50–80.

10 Beyond the enclave

Workforce mobility and livelihoods in a New Caledonia mining region

Séverine Bouard and Valentine Boudjema[1]

Introduction

Mining development has the potential to transform household activities. Numerous studies have found that in Oceania and elsewhere in the world, the ratio of time invested in agricultural and hunting/fishing activities decreases, while increased participation in non-agricultural activities (e.g., mining-related employment, setting up and/or working in local businesses, renting out property) creates new resources and alters the distribution of household assets and conditions of socialization (Gaillard 2010; Sabinot and Lacombe 2015; Sabinot and Herrenschmidt 2020). Livelihoods also seem to change in territories not directly impacted by mining activities – territories that we refer to as the mining "margins," which often provide mobile employees. The mine's impact thus goes beyond the area of the extractive activity, due to spatial mobility and family networks (Castillo and Brereton 2018). Hence, it is important to include the margins beyond the boundaries of mining concessions in analyses of the transformations driven by mining and metallurgical activities (Bainton et al. 2017), through attention to flows not only of labor but also of materials and ideas when employees return home or when mining companies invest in CSR programs in these more remote territories (Hinojosa 2013; Brugger and Zanetti 2020; Dessertine 2021). By using the case study of the North Province of New Caledonia – namely the mining territory of the Voh-Koné-Pouembout-Poya (VKPP) municipalities and their connections to the Hienghène mining margin – this chapter highlights the socio-spatial changes brought about by mining and its associated mobilities and points to how the changes observed call into question the concept of "mining enclave."

New Caledonia is rich in nickel ore and has long been exploited, even before the twentieth century. The nickel sector currently directly employs almost 6,000 people, some nine percent of private sector employment (Desmazures and Laloum 2021).[2] In the North Province, the majority is pro-independence, and the development of the nickel sector is at the heart of its strategy for achieving full sovereignty. The three provinces – the North, the South, and the Loyalty Islands – have been the competent authorities in the mining and economic development fields since the Nouméa Accord in 1998, and plan to finance their public policies of territorial rebalancing[3] and economic diversification from nickel sector economic spin-offs. The fact that these authorities establish nickel as one of the pillars of the country's

DOI: 10.4324/9781003313236-13

economy and that the resource-free areas are also those with the most unequal living conditions nationwide would suggest a spatial approach to mining territory and its margins.

This chapter analyzes the way mining territories and margins transform each other in the North Province, particularly via the circulation of employees of the Koniambo Nickel SAS (KNS) mining and metallurgical company. The first section looks at how to define "mining territory" and the concepts of "margin" and "mining enclave." It also presents our approach, which combines qualitative and quantitative methods. The second section describes the case study of the North plant in the VKPP mining territory, and Hienghène, a municipality on the North Province's eastern coast that we describe as a mining margin. The study addresses VKPP and Hienghène together to make a critical assessment of the mining enclave concept. The third section focuses on the first two phases of the mining and metallurgical cycle (i.e., the KNS smelter construction and entry into production). It analyzes how the company's search for a "social license to mine" shaped mobilities and made the mine an integral part of people's individual lives. The fourth and final section explores the impacts of the employee mobility described by observing changes in household activities and livelihoods. The descriptions of employees' "return to the margins" help explore further what mining work "does" to workers and territories, to shed light on Kanak individuals' links to and engagement with their territories of departure (the margins), and on their perception of these territories, including the way they project their life experiences (employment at KNS) into these images and feelings.

Results are based on three main corpuses of data: (a) a long-term ethnography by Séverine Bouard, who since 2006 has been living and working in Koné, a municipality in the North Province of New Caledonia; (b) two surveys[4] she has conducted with the local research institute IAC (New Caledonian Agronomic Institute) and CIRAD (French Agricultural Research Centre for International Development) since 2011 using a diachronic approach (Guyard et al. 2013; Bouard, Apithy, and Guyard 2018; Sourisseau et al. 2021), and qualitative studies on social change related to the development of Kanak women's participation in nickel mining employment[5] (Mazer, Mills, and Bouard 2022; Vadot et al. forthcoming); and (c) 35 interviews conducted by Valentine Boudjema for her PhD fieldwork (2020–2023) on material and non-material circulation between the mining territory and its margins in New Caledonia.

Conceptualizing mining margins

This chapter is based on the analysis of the socio-spatialization of mining and metallurgical activity. How should a "mining territory" be defined? Is it the mining concession? The municipality that hosts the extractive activity? Or should the definition integrate other territories which may be labor reservoirs and connected to the mining territory by flows of ideas, skills, practices, and money? We take a twofold approach by considering both the mining territory and its mining margins, the latter being areas that have no mines or processing plants, but which are linked to them by material and non-material flows. Indeed, mining territories and their margins are constantly made, negotiated, and renegotiated by processes and flows that connect places through mobility of money, things, people, and ideas (Bebbington and Humphreys Bebbington 2018). This constant molding of space is the guiding thread of the chapter, which questions the idea of a mining enclave in the North Province of New Caledonia (Le Meur 2021).

As it is typically conceived, a mining enclave is thought to be illustrated by the immunizing posture of the mining company (Donner 2011) – i.e., isolated from its territory and acting without it – which both appropriates land and concentrates capital (Levacher 2016). This concept sees mining and metallurgical centers as disconnected from their local contexts, operating as autonomous eco-spatial units (Ferguson 2005; Magrin 2013; Filer and Le Meur 2017). Our theoretical positioning is supplemented by a time-based approach that considers the plant's construction and production stages. This reveals constantly changing networks and flows, proof of a continuously renewed link between the sub-spaces of the mining territory and its margins, and thereby invalidates the hypothesis of a mining enclave around the North plant.

Connecting to the mining territory

Mobility and flows during the metallurgical cycle of the North plant

The North Province is characterized by intense flows of employees, skills, and social representations associated with KNS's operations – known as the North plant (Vavouto industrial site and the Koniambo mine) though also commonly called the "country's plant" due to a majority public ownership strategy of the North Province, with a 51 percent/49 percent shareholding model. The area of the VKPP municipalities (the VKPP urban area), understood here to be the mining territory, is notable for several reasons (Figure 10.1). It hosts the third and last nickel processing plant built in the country. Historically, the KNS plant is, for the North Province, the symbol of the rebalancing expected of the Matignon agreements (Carnuccini and Guillaud 1999; David, Guillaud, and Pillon 1999; Kowasch 2010; Pestaña, Bouard, and Sourisseau 2013). The construction of the North plant began in 2008 and has transformed the territory by introducing new forms of economic governance (Le Meur et al. 2012; Grochain 2013). More broadly, the exploitation of the Koniambo massif, the construction of the plant, and the ensuing urban and demographic development of the Koné municipality (Kowasch 2010; Dumas 2020; Lion 2022) have led to social and spatial changes that can be observed by analyzing mobility and livelihoods. On the eastern coast of New Caledonia, the Hienghène municipality is the extreme opposite of VKPP and is an archetypal case study of the mining territory (Figure 10.1). The VKPP mining territory and the Hienghène margin make up the Hoot ma Whaap Customary Area, a cultural dimension of the territory characterized by flows of subsistence practices and kinship ties which, as discussed below, facilitate the employment of inhabitants of the margin (Hienghène) at the North plant.

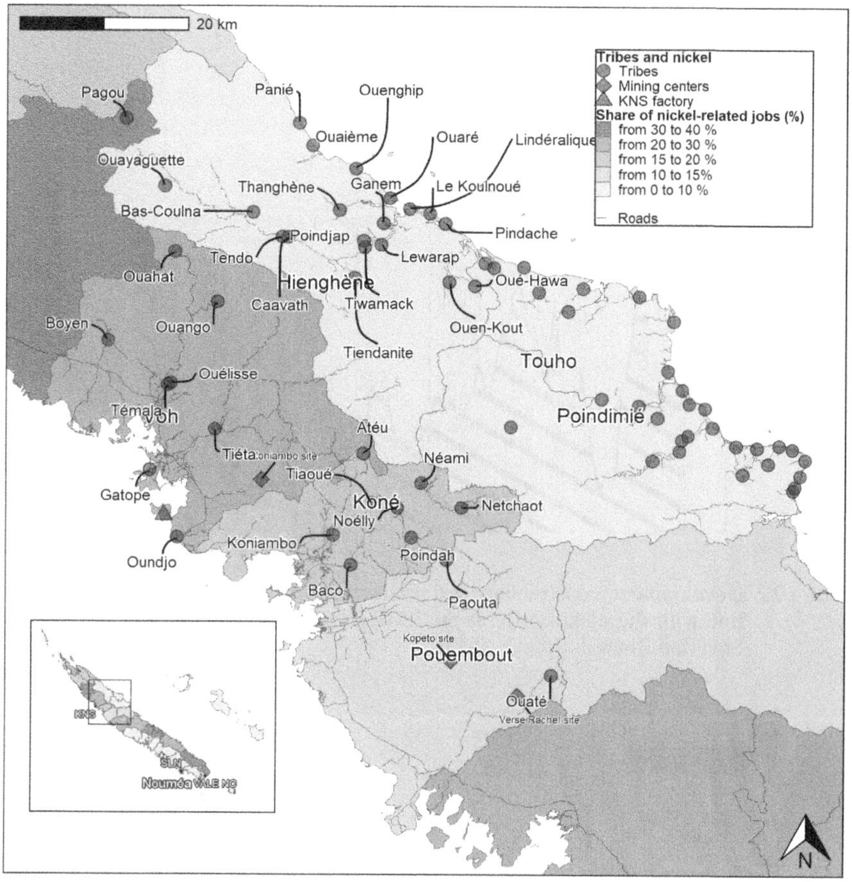

Figure 10.1 Voh, Koné, Pouembout, and the East Coast showing the share of nickel-related jobs. Map by Jonas Brouillon. Source data: ISEE 2019.

This section describes the circulation of employees according to the stages of the mining cycle. During the nickel plant construction phase, there was an intense flow of people from the margins toward the mining territory, mainly to work in the construction trades, with two reverse flows during demobilizations (i.e., the end of employment contracts) in 2008 and 2010 (Figure 10.2).

In these periods of reduced employment opportunities, some people stayed and became employed in the territory in both mining and non-mining jobs, while others returned to the margins. Data from the four general censuses conducted between 2004 and 2019 by the ISEE (2019) indicate that the number of people from the North Province's municipalities who migrate to VKPP has fluctuated, with Pouembout and Koné being the two municipalities that are experiencing a clear demographic expansion compared to the municipalities of Voh and Poya (Figure 10.3).

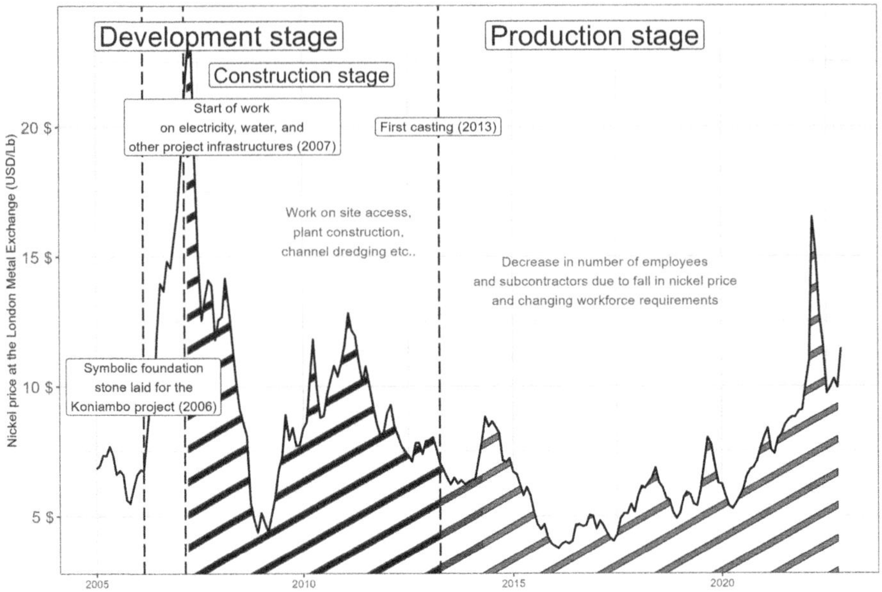

Figure 10.2 Development, construction, and productions stages of KNS and its impacts, in line with the nickel price at the LME between 2005 and 2021. Designed by Séverine Bouard. Source: ISEE.

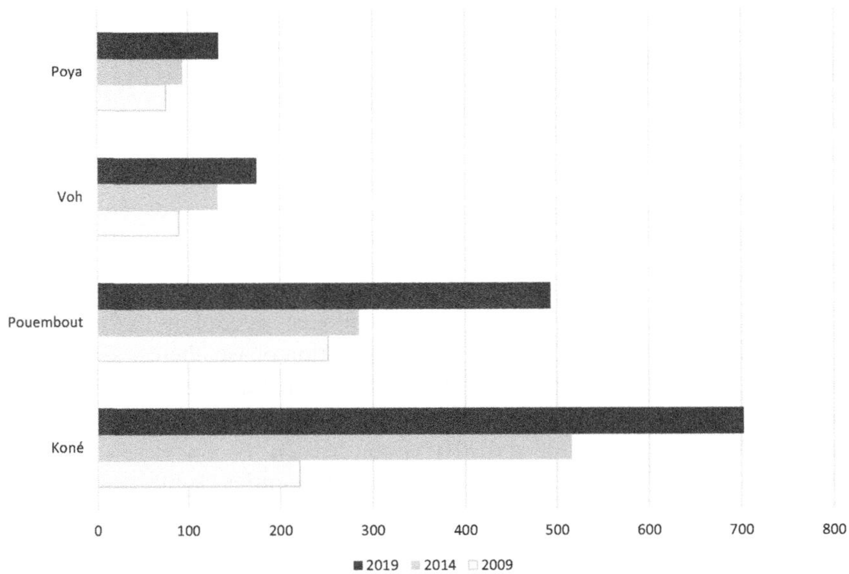

Figure 10.3 Number of individuals from non-VKPP parts of the North Province who moved to the VKPP zone between 2009 and 2019. Source: ISEE, migrations internes, commune antérieure. Unit: individual. Interpretation: In 2009, 90 people from the North Province settled in the municipality of Voh, 252 people arrived in Koné, 221 in Pouembout, and 76 in Poya. These figures are reached by adding together all arrivals from all the municipalities of the North Province, the Bélep Islands excluded.

Kanak people interviewed over the construction stage (Figure 10.2) testified to a sharp increase in some families' standard of living, which was visible in customary ceremonies, particularly weddings. Although this opulence did not last, and interviewees described a return to normality once the preliminary work on the plant was completed, expenditures for customary ceremonies were discussed by those interviewed as excessive and growing exponentially from one wedding to the next. Nonetheless, a clan or family would typically show affluence to increase its social and customary status, which illustrates Gregory's (1982, 1997) points about exaggerated ceremonial exchanges in an expanding market.

At the same time, an international migratory dynamic appeared. To create a sufficient local employment pool, then mining partner (Falconbridge) and the North Province subsidized training programs in Canada and France for North Province residents (Salaün 2014), in return for a commitment to work at the North plant after training. Despite this proactive approach, results remained mixed. The plant was behind schedule, and it was sometimes hard for returning New Caledonians trained abroad to get hired (Gorohouna 2021). KNS later acknowledged that the training programs were not sufficient, in part because the number of North Province nationals trained for employment in management positions remained limited (Kowasch 2010).

The North plant gradually entered into production in 2013, and demand for metallurgy and extraction workers changed with the need for a more skilled workforce. The employment pool was widened to include the whole country and even beyond (due to lack of training in the territory, see above). The production phase had high employee turnover, which contributed to increased mining-employee traffic within the North Province and between the North Province and the Loyalty Islands Province. This has been counterbalanced by KNS's "snail-shell shape" subcontracting policy negotiated in 2002 with Kanak villages, also called "tribes" by interviewees, around the mining massif, for whom employment was a priority. The company has since helped create several Simplified Action Companies (SAS), subcontracting networks that foster local entrepreneurship (Grochain 2013).

The "snail-shell shape" policy and its consequences on individuals' mobility

In 2020, the KNS external affairs director reported, "We have a snail-shell shape policy when it comes to subcontracting." Here, the snail shell is a metaphor for KNS's socio-spatial recruitment strategy for subcontractors and employees at the mine and smelter (see Figure 10.4). The importance the strategy places on spatial proximity is not simply a company decision. The negotiations between Kanak landowners and KNS during the construction of the plant produced a recruitment strategy that prioritized people from tribes close to the massif, before individuals from the Voh municipality, the rest of the VKPP municipalities, the North Province, New Caledonia as a whole, and finally, foreign countries. The landowners of the mining massif and the surrounding tribes are at or near the center of the snail-shell spiral, while foreign countries and expatriate workers are at its outer edge.

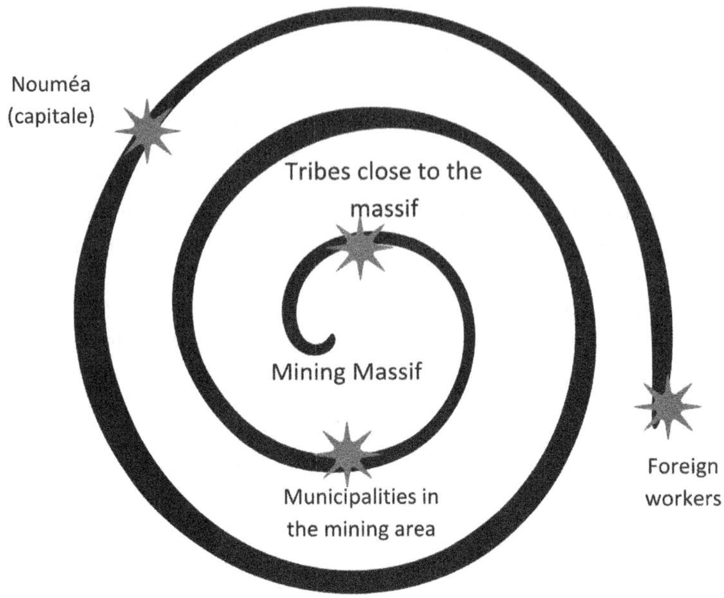

Figure 10.4 The snail-shell shape policy of recruitment. Figure by Valentine Boudjema.

The snail-shell shape policy reflects two related issues: local demands for employment and the social acceptance of a mining and metallurgical project. As with any such project, KNS must ensure that it has a "social license to mine" (Owen and Kemp 2017). The 2002 agreement with Kanak landowners was the first step in building the social acceptability of the North plant, which must be continuously maintained. Negotiations conducted by the Kanak customary authorities in the Koniambo mining territory demonstrate the link between local demands for employment and the need for the company to secure its social acceptability, with KNS outlining and justifying in its annual social balance sheets the evolution and the balance between local and external employment in its total workforce. KNS has been progressively working to increase its local workforce by developing an organized network of subcontractors and implementing training schemes for youth from the tribes, to fulfill past agreements (Gorohouna 2021).

A 2010 local employment law further supported the local recruitment strategy in KNS's 2002 agreement with the Kanak landowners. The law was enacted for the private sector after lengthy debate and gives New Caledonian citizens[6] priority in local labor markets. During the KNS North plant construction, the North Province – which views the project as a mechanism for economic, spatial, demographic, and ethnic rebalancing across New Caledonia – plans for a spatial labor recruitment strategy to the benefit of its constituency (i.e., North Province residents). The preference that KNS gives to subcontractors from the landowning clans of nearby tribes demonstrates the power of political and customary authorities in negotiating economic benefits from the mining project, which has undoubtedly further shaped mobility dynamics in the case study area.

So how has the "snail-shell shape" policy influenced people's mobility and flows? Has it led to less recruitment in the mining margins, such as Hienghène, to the advantage of job seekers and subcontracting companies in Voh and VKPP? In 2020, KNS's workforce was as follows: of the 1,086 employees, 458 employees (42.2 percent) said they were from a North Province municipality, of which 158 were from VKPP and 53 from Voh (14.5 and 4.9 percent of total KNS employees). In other words, of the total number of KNS employees from the North Province, 34.5 percent came from VKPP, the mining territory. Thirty-one workers said they were from Hienghène (three percent of total KNS employees). Interviews with company employees suggested that KNS has worked to train people from VKPP municipalities but is struggling to keep them. As an environmental technician put it, although "the company wants to have people who know that this weekend they will be at home here resting [i.e., living in the mining area]," it is faced with "the wishes of the people here. What do they want to do? Do they want to stay?. . . The people here do not want to work in the mine." Revealingly, the local expression "mining" means to go and earn fast cash in the sector.

Impacts of employment at KNS on individual trajectories: from entry to circulation

In addition to highlighting employee mobility between territories, this study reveals the different forms of capital (human, financial, social, etc.) employees acquire in the process, and that transit with them. Surveys conducted in 2018 of women working at KNS (Vadot et al. forthcoming), including women from Hienghène, reveal two main pathways to employment at the mine or smelter, pathways also shared by men workers. The first is education in the geosciences, which is available at undergraduate level in New Caledonia, although graduate level is typically completed internationally. The second is the Adapted Military Service Regiment, which is for young people (including those failing at school) to acquire general knowledge in promising trades, and particularly to obtain driving licenses for light vehicles. Few new workers in the mining and metallurgy sector require significant retraining, as employees tend to come from the construction and government infrastructure sectors, have skills similar to those needed in mining and nickel processing, and machines are exchangeable between the two sectors (Bassuel 2013; Briand 2021).

Employees aged over 50 often say they want to work in the mining and metallurgy sector in political terms (e.g., to work for the "country's plant"). However, this identity-based motivation is fading. For younger workers, the North plant is "the fight of the elders," meaning that although the older generation's quest for independence through nickel is important, it is not the main reason why they look for jobs at KNS. The generation that viewed the success of KNS as key to nationalism is not necessarily the one that works there today. In the words of Sonia, who has worked in KNS's human resources department for many years, "There isn't the same commitment. . . . When people used to come to Koniambo nickel, they were really proud to work for their country's beautiful project."

The study also looks at employees for whom KNS is a second choice because they have not found work in their own field, but wish to be active and improve their standard of living. All the employees interviewed said they wanted financial independence, particularly the women (Vadot et al. forthcoming). Employment at the mine can be a break for people who later resume life in the tribe. Many young men seek a one-off, short-term job at the mine and then return to the tribe, often leaving again to find a job outside the tribe for money reasons or just because they want to.

However, interviewees often reported that while employed at KNS they moved weekly, or at least monthly, between the mining territory and their home territories, including those in the mining margin. The four-days-on-and-four-days-off schedule allows employees to return to Hienghène or other municipalities frequently, particularly on the eastern coast, although the trips are tiring. Returning to the

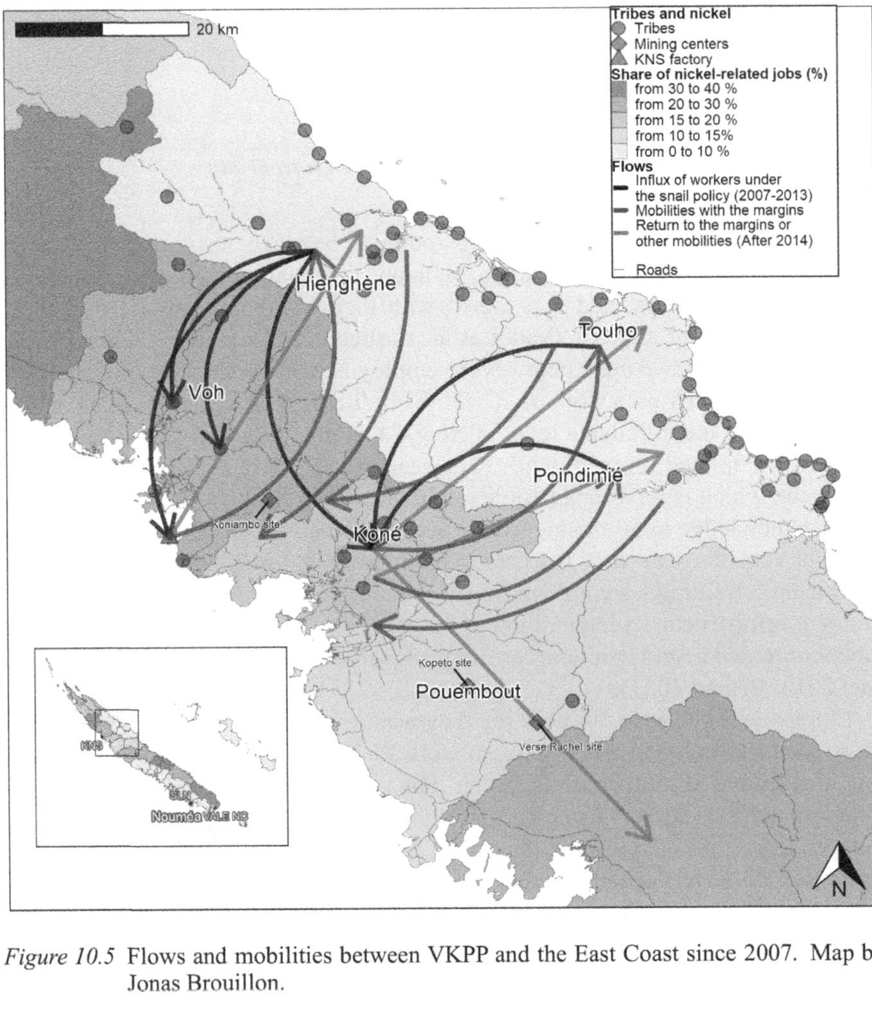

Figure 10.5 Flows and mobilities between VKPP and the East Coast since 2007. Map by Jonas Brouillon.

margins for a while is a relief from the pressure of achieving KNS production goals, and interviewees often referred to these return trips as "recharging their batteries." For many, even if their homes are several hours away by road, being able to return to them regularly is fundamental to their well-being.

In sum, the construction of the "country's plant" in the North Province has led to the creation, transformation, or disappearance of worker circulation trajectories and associated flows (Figure 10.5). The mining cycle and KNS's search for social acceptability are variables that have influenced worker mobility within the mining territory and between it and the mining margin. Snail shell-like mining micropolitics promotes a local employment pool restricted to the tribes around the massif. One might think that this micropolitics would, in effect, isolate the mine and plant from the rest of New Caledonia, with most employees coming from the closest tribes and villages. However, due to the lack of training in the territory and high employee turnover, KNS is expanding its employment pool to the near and distant margins. The distant margins are the Loyalty Islands and are not considered in this chapter. Flows, including the material flow of employees, question the idea of enclave around the Koniambo plant. Interviewees' references to the "country's plant" show how the motivations for employee circulation have changed. Younger generations describe the plant as the "fight of the elders," but for them, it primarily represents employment and financial independence.

Back and forth mobility, changes in the mining territory and in the margins

Focusing on the skills that people mobilized, as acquired intangible capital, makes it possible to better understand the back-and-forth mobility observed among KNS employees and subcontractors. Here, we explore the evolution of livelihoods in the mining territory (VKPP) in the construction stage (when people arrived in the mining territory, 2006–2014) and the production stage (2014-to date), to demonstrate the evolution of household capital in these territories. By combining quantitative and qualitative approaches, the authors capture both material flows and the harder-to-capture non-material ones.

Pluri-activity: adaptation and resilience of farming activities

Household-level analyses confirm the prevalence of pluri-activity in the VKPP mining territory, even in quite remote tribes. On average, households in the 2018 sample reported 3.6 types of activity. In addition to agriculture and livestock farming, wage labor is a source of income for nearly 80 percent of tribal households and is essential to household reproduction. Agricultural and livestock systems are flexible and enable people to respond, individually and collectively, to wage-earning opportunities, including the regular circulation of people between KNS, the village, and the tribes (see above).

Pluri-activity in part spreads economic risk, where permanent jobs are scarce, and the Kanak population remains less qualified and less well-paid overall than the rest of the population. In an economy still dependent on government transfers (which distort domestic prices) and dominated by non-market services (Cerom 2008; Mathieu, Couharde, and Pestaña 2016), wage employment guarantees higher income levels, and it is therefore not surprising that it dominates the aspirations of Kanak people. However, pluri-activity is also based on a social logic and life choices. Farming, having access to tradable sea or meat products, working and not just residing in the tribe: these activities are all part of an identity that anchors family groups in their clan and tribe, in addition to being an "art of living." They also promote community solidarity, including through practices of gift exchange, whether of equal value or to mark social hierarchies (e.g., between generations). Gina, a truck driver at KNS, explained in an interview in 2018 how and why she regularly travels between VKPP and the eastern coast and continues to farm there:

G: Well, I come back, I do my yam field. After that, we don't really need to take care of it. . . . At the end of the year, one field is always left for food, and another for custom, because yams are part of the custom. So we keep it to participate in the custom.
M: For weddings and funerals.
G: That's it – deaths, weddings, and everything. There, we had three weddings in a row. Well, we gave our share. Now we are redoing our field, so that we can give again next year. . . .
M: But how often do you go back?
G: About twice, three times a month.

Mining jobs also provide skills that connect workers to the margins. The Hienghène municipality recently hired former KNS employees in the town hall technical department because they can drive heavy vehicles for road and ditch maintenance. The technical services director recognizes that they are responsible and respect schedules and regular work hours, which he attributes to their work experience at KNS. A second example is the spirit of initiative and the self-confidence that leads to entrepreneurship. A KNS technician explained that he learned management and accounting skills that are useful for entrepreneurship and is very confident in his new abilities. When skilled workers go back to the margins, they mainly work in agricultural activities, or in one case, a real estate activity (house rentals). This usually involves men over 50 who wish to return permanently to their tribe and who have accumulated experience and sometimes money.

Wage employment in the mine and the mobility entailed increases individuals' capacities because they acquire different capital. The section above shows that skills are an important (human) capital that moves from the mining territory to the margins. This intangible capital includes the networks and social capital of workers from the margins that expand and are strengthened when workers arrive in VKPP. For some people from the margins but with customary ties to clans in VKPP, employment-related mobility brings families closer together. Participating in customs, having coffee with the family, or taking a few hours to fish or hunt strengthens family networks in the mining territory.

A third kind of capital mentioned by interviewees above all others is financial, as illustrated by this extract from an interview with a CSR officer at KNS from Hienghène: "Those who leave here, yeah, they have power, power through income, the income from the mine. . . . If you haven't invested here [VKPP] you will have invested in Hienghène. If you can invest in both, so much the better." The officer indicates that money is a flow and acquired form of capital that can, in turn, deliver land and social capital, the "power" he refers to, acquired through income from mining employment. Money sent to the margins can enable people to take part in customary activities. A wedding, for instance, can be the most significant expenditure in an employee's household budget. The CSR official also pointed to the direction of the financial flow, as wage earners may invest in the mining territory and in their tribe of origin (in Hienghène).

Financial capital produces a fourth, physical capital, such as a car or a house. There are diverse strategies for investing in a house, but unvaryingly the clan house, called the "family house," is the first to be renovated and, sometimes, extended. Next is the parents' house. Finally, sons who will not inherit the parents' house will build their own. Those who inherit houses and do not have to build their own in Hienghène, or those who have the means (e.g., eldest siblings or those in positions of responsibility), often choose to build or buy a house in the mining territory. This investment illustrates their dual location: they work in the mining territory and rest and carry out their status and customary duties in the margins. Hienghène remains the place for rest, weekends, holidays, and retirement. But what happens in the mining territory when activity slows, after the boom of the plant's construction phase?

High vulnerability of off-farm incomes: shrinking salaries post construction stage

Our diachronic approach allows us to analyze changes in the relative significance of wage income in livelihood systems. Wage employment is dominant in people's livelihoods, especially for tribes in the VKPP mining territory, and income levels from wage employment are significant. In 2018, average annual salaries were nearly 22,700 USD in the 77 households surveyed. Regarding entrepreneurship, only six households living in tribes stated that they had non-agricultural income from their own business in 2018 (Sourisseau et al. 2021). Clearly, even if the VKPP area encouraged entrepreneurship during the development and construction of KNS's North plant (2008–2014), this has changed since 2016 (Figure 10.2).

These changes in the mining territories affect household incomes. In the eight years spanning the construction and production stages, the average annual cash income of households surveyed fell by 12 percent, from 32,000 to 28,300 USD. The decline in wage income alone accounts for more than 80 percent of the income gap between 2010 and 2018 (Sourisseau et al. 2021). After the favorable economic climate of 2010 – arising from the 2008 nickel boom, the North plant construction and its spin-off effects – economic growth in the area has slowed considerably, and with it employment opportunities (Figure 10.2). Some young workers hired during plant construction, or during its first years of operations, were laid off when the price of nickel fell in 2015. Some of them then chose or had to return, temporarily or permanently, to the territories on the margins, where they took advantage of assets developed in the boom period. Nevertheless, reduced household income is also explained by a de-cohabitation process – when people from the same household stop living under the same roof. Women workers especially move from the tribes to the village, looking for more neutral gender roles than those of the tribe.

Overall, these changes do not dramatically reduce the importance of cash income for households in VKPP but do give greater strategic importance to other activities. We also observed in survey results an increasing dependence on state social transfers (up 71 percent between 2010 and 2018) in the 77 households followed.

Family farming, fishing, and hunting still prevalent to maintain social capital

In 2010–2018, parallel to the contraction in wages, households became much more involved in farming for both consumption and for sale (incomes from agriculture increased by 115 percent), reinforcing agriculture's place in livelihood systems. These trends reflect two essential features of Kanak livelihoods: the preponderance of pluri-activity and the significance of non-market exchanges for agriculture, fishing, and hunting products. These two features, in turn, point to the role of human and social capital in obtaining monetary income. Belonging to knowledge and support networks is helpful in finding employment, and households seek to maintain these networks by donating products and participating in dense gift exchanges. Falling wages have led to a ten percent decline in average total income, and households further adjust their monetary and non-monetary livelihood activities by capitalizing on the size of their fields to sell food crops, by searching for off-farm employment, or by relying on government support via social transfers.

Return to the margins: what working at the mine "does" to Kanak people and territories – imaginaries and practices

The return to the margins poses questions about people's link with the tribe and identity, particularly in their relationships with nature. In the light of the descriptions above of employee mobility and other associated flows between the mining territory and the margins, this last section analyzes how these affect symbolic dimensions of territory. We are interested in the impacts of people's work experiences at the mine or plant – and of related mobility practices – on workers' representations of both mining and non-mining territories, and on their farming and subsistence practices. Working at a mine or a plant is a dramatic change in landscape, pace of life, consumption practices, and social relations. The soil in the mining area is red. There is dust and asbestos. Workers must wear masks, and the truck cabs need to be regularly cleaned. It is noisy and schedules are hard, as workers often alternate between day and night shifts. Although employees are generally enthusiastic about their jobs, they describe them as hard. Interviewees – especially those who no longer worked at KNS – often contrasted this exhausting environment with that of the tribe, where livelihoods are rooted in agriculture, fishing, hunting, and small businesses, along with jobs in the local administration, tourism, or education. In some interviews, Hienghène was described by inhabitants as the "green mine." This image reflects a social representation of the mine as wealth, but insists that Hienghène possesses other valuable resources, which are not the red ones of the mine and nickel.

Employees dissociate the two territories temporally, with the mining territory associated with the short and medium terms of work and career, and the margins with the long term of family, investment in customary practices, preparation for retirement, and anchoring culture and identity through farming and customary status. The dissociation in terms of subsistence activities depends on the individual's standard of living and salary. The higher the mine worker's socio-professional category, the higher the salary, and the greater the likelihood that they establish a second residence in the mining territory. In these cases, the employee may own two fields, one in the mining territory of VKPP and one in the margins (i.e., non-mining area), with the larger one in VKPP. One example is Nadia, a plant operator who has an agricultural field with her husband on the eastern coast in Poindimié, and who has just invested in a house in Pouembout.

The margin is therefore the long-term territory – the one that makes a return possible, because food resources and family are still there. Individuals from Hienghène indicate that they still grow food, despite having entered the wage-earning sector at some point. John, a truck driver at KNS, stated in a 2022 interview, "It's hard to say why I farm, because I farm a small plot, just to keep this link to the land." John said, "the link to the land" was why he kept his field going, even though he was making it smaller and did not do all the work alone. Instead, he sometimes calls on family labor, and the resumption of this mutual aid network is undoubtedly as important as the fact that he also farms the land himself and brings in the fruit of his labor. These mutual aid relationships, whether paid in kind or free of charge, are fundamental to reintegrating and "reembedding" mining employees on a daily basis. In the end, non-market, domestic-economy activities, as shown in statistics quoted above (see also Bouard, Apithy, and Guyard 2018), remain attractive to Kanak men and women.

KNS workers, even more than their managers, explain that their "diversion" to the mine will not last long. By remaining active in the two territories, they accumulate two forms of work: work in the mine and work in the tribe, and often value the latter more highly. Even while employed at KNS, employees continue to invest more money and time in the margins than in the mining territory for fishing and farming. Although interviewees explained their relationships to territories (mining and non-mining) as different, complementary, and articulated, they were clearly aware that nickel stocks are limited. They noted that they can always return home and live off their food crops. They see agriculture and the valorization of natural capital through farming as a means of subsistence: along with harvesting, agriculture is a safe haven in the event of the loss of mining employment. In VKPP, surveys show that the surrounding tribes continue farming and show a slight increase in cash income from agriculture after a contraction in mining employment, which further illustrates this phenomenon within the mining territory itself.

Nevertheless, this must be qualified: although agriculture, fishing, and forestry resources are always there, those who live in the margins recognize that the safety they offer has limits. In particular, they are aware of the narrowness of agricultural markets in these areas, which are far from urban centers, and of the difficulty of transforming a food-producing agricultural activity into a remunerative activity (e.g., marketing their products). They also realize they need know-how to do so.

Hence, when people return to and resume social relations in the margin, it is not just to (re)develop land but also to regain their place in the local social hierarchy, through customary wedding practices for instance, and/or by officially becoming a clan leader (a position for men). Customary social relations have remained unbroken. The Voh and Hienghène municipalities, for example, belong to the same customary area, called *Hoot ma Whaap* in Fwâi, a Kanak language. This linguistic and cultural area, through its material and non-material flows, integrates and links the VKPP mining territory and the Hienghène mining margin. These connections exist through the customary paths, connections, and alliances that unite the Voh and Hienghène clans. Besides these non-material flows, forms of social capital forged through hunting in the mountains and through family networks help individuals from Hienghène circulate to work in KNS, in Voh. Pre-existing food-harvesting practices and marriage alliances continue and are transformed by the North plant project. However, the trajectories of several individuals from Hienghène and other east coast municipalities demonstrate the importance for people of re-establishing themselves in their place of origin within the customary hierarchy; and of *suspending* movement between territories in order to be a full-time clan chief. Hence, even though returning to the margin is not always easy in economic terms, people sometimes have to resign from their wage-earning job at KNS in order to fulfill the customary duties of clan chiefs.

Finally, some employees choose to return to life in the margin completely because the relationship of mining with nature becomes unbearable. As Graziella, former manager of a mining revegetation company, a subcontractor for KNS, explains

> I stopped mining. Why? . . . At a certain point, KNS put the environment aside. That was the moment when I said, 'No, stop.' I don't support things like that anymore. . . . They stopped funding the environment. They financed a lot to avoid maximum pollution, but in terms of the environment, they tore down trees, but they didn't replant them; they covered them up. They. . . that's it. I said, 'No, I'm stopping here.'

For people like Graziella, returning to the margin is no longer linked to working conditions *per se*, but to a refusal to participate in damaging nature. The mining company's destructive relationship with nature shapes the job choices of some employees, who refuse to drive shovels and "dig up the mountain" and ask just to drive trucks, or who look for jobs in environmental divisions (e.g., water management or revegetation). But this "negotiated" position is often temporary, and when it becomes untenable, employees choose to return to the margins to rebuild a relationship with nature that is not detrimental to it.

Conclusion

In interviews, plant employees regularly spoke of the "confrontation of two worlds." They mean the worlds of the tribe and of the company that are molded through the mining territory and its margins. This chapter analyzes what we call the near margin, that of Hienghène, a three-hour drive from the North plant. This band of non-mining areas shows that it is not effective to restrict the concept of the mining territory to the North plant, as there is considerable circulation between areas and they connect the territories, so the mining territory and the margins become one single area polarized by the mining and metallurgical enterprise. Relations between the two sub-spaces have evolved over time but have always existed. Today, relations and circulations have multiplied and intensified, demonstrating a single mining system where the mine transforms but also maintains the margin. The value ascribed to each place, through social representations and emotion, feelings, and sense experience, demonstrates that the margin must exist for the Kanak to work in the mine. This territory provides the balance sought between the financial and family emancipation obtained on the mining territory and the link to nature and to the clan, the identity value, reinforced by the margin. This analysis suggests the need to rethink the concept of the North plant as a mining enclave. The mining and metallurgical area is not isolated and partitioned, and borders are not hermetic: whether through the expansion of the employment pool to the margins, or through the circulation of skills, equipment, money, ideas, and practices, the North plant connects to the territory of the North province as a whole.

Notes

1 As the authors contributed equally, names are listed in alphabetical order.
2 In comparison, the tertiary sector (services and retail) concentrates 66 percent of private jobs (43,000 individuals) and agriculture accounts for 2.7 percent of the workforce (1,800 individuals) (Ardoino et al. 2021, 73).
3 The notion of rebalancing derives from the strong spatial and economic imbalance of the territory, accentuated during the colonial period with its centralizing policy. Noumea, located in the non-independence South Province, captures most of the population and economic activity, creating a marked imbalance with the rest of the archipelago. Kanak people represent 95 percent of the Loyalty population, 72 percent of the North Province and 29 percent of the South Province. In 2015, the median income in the South Province was twice that of the North Province and two and a half times higher than in the Loyalty Islands Province (Gorohouna and Ris 2015). The North Province shows an over-representation of individuals with the lowest standard of living (Roussineau 2020).
4 "Agriculture in tribe, Kanak village" conducted in 2011 (Guyard et al. 2013; Bouard, Apithy and Guyard 2018) and RACINE (Sourisseau et al. 2021) conducted in 2019 in VKPP. In the second study, 77 households living "in tribe" on customary land (Kanak) were surveyed for the year 2018 and had already been surveyed in 2010. This diachronic approach allowed us to understand changes that occurred during the two first stages of the mining cycle (construction and development) in the mining territory.

5 Qualitative studies financed by the research program MINERAL (The Knowledge Network Mining Encounters and Indigenous Sustainable Livelihoods: Cross-Perspectives from the Circumpolar North and Melanesia/Australia). See: https://www.mineral.ulaval.ca/en/about/objectives-and-research-themes

6 The Nouméa Accord established a New Caledonian citizenship, based on a restricted electoral body for local elections and for the referendums on self-determination.

References

Ardoino, Magali, Yann Caron, David Chatelain, Baptiste Dagorn, Elodie Pichon, and Andréa Tran Van Hong. 2021. *Rapport annuel économique Nouvelle-Calédonie 2020.* Nouméa-Paris: IEOM.

Bainton, Nicholas, Vlado Vivoda, Deanna Kemp, John Owen, and Julia Keenan. 2017. *Project-Induced In-Migration and Large-Scale Mining: A Scoping Study*. Brisbane: Centre for Social Responsibility in Mining (CSRM); University of Queensland.

Bassuel, Anna. 2013. "L'utilisation des engins de BTP: reflet de l'agriculture néo-calédonienne." Master diss. IAC. Montpellier SupAgro.

Bebbington, Anthony, and Denise Humphreys Bebbington. 2018. "Mining, Movements and Sustainable Development: Concepts for a Framework." *Sustainable Development* 18 (5): 441–449.

Bouard, Séverine, Leïla Apithy, and Stéphane Guyard. 2018. "Family Farming in Contemporary Kanak Society." In *Diversity of Family Farming Around the World, Existence, Transformations and Possible Futures of Family Farms,* edited by Pierre-Marie Bosc, Jean-Michel Sourisseau, Philippe Bonnal, Pierre Gasselin, Élodie Valette, and Jean-François Bélières, 285–296. Versailles: Quae Springer International Publishing.

Briand, Antoine. 2021. "Le développement d'une commune à la marge de l'activité minière et métallurgique: le cas de Bourail en Nouvelle-Calédonie." Master diss. ISTOM. Angers.

Brugger, Fritz, and Jessica Zanetti. 2020. "'In My Village, Everyone Uses the Tractor' Gold Mining, Agriculture and Social Transformation in Rural Burkina Faso." *The Extractive Industries and Society* 7 (3): 940–953. doi: 10.1016/j.exis.2020.06.003.

Carnuccini, Anne-Sophie, and Dominique Guillaud. 1999. "Les fumées de l'usine: un projet métallurgique à l'appui du rééquilibrage." In *La Nouvelle-Calédonie à la croisée des chemins : 1989–1997*, edited by Gilbert David, Dominique Guillaud, and Patrick Pillon, 83–99. Paris: IRD - La Société des Océanistes.

Castillo, Gerardo, and David Brereton. 2018. "The Country and the City: Mobility Dynamics in Mining Regions." *Extractive Industries and Society* 5 (2): 307–316. doi: 10.1016/j.exis.2018.02.009.

CEROM. 2008. *Les défis de la croissance calédonienne*. Nouméa: CEROM.

David, Gilbert, Dominique Guillaud, and Patrick Pillon. 1999. *La Nouvelle-Calédonie à la croisée des chemins: 1989-1997*. Paris: IRD - La Société des Océanistes.

Desmazures, Elise, and Maurice Laloum. 2021. "L'impact du nickel en Nouvelle-Calédonie. Les emplois directs et indirects." *ISEE* 47: 6 pages.

Dessertine, Anna. 2021. *À quoi tient le village. Espaces et mobilités en pays malinké (Guinée).* Nanterre: Société d'ethnologie, coll. "Sociétés africaines."

Donner, Nicholas. 2011. "Notes sur la dimension immunitaire des enclaves pétrolières." *EchoGéo* 17. doi: 10.4000/echogeo.12555.

Dumas, Pascal. 2020. "Challenges and Development Prospects in New Caledonia." *Information Géographique* 84 (1): 36–58. doi: 10.3917/lig.841.0036.

Ferguson, James. 2005. "Seeing Like an Oil Company: Space, Security, and Global Capital in Neoliberal Africa." *American Anthropologist* 107 (3): 377–382. doi: 10.1525/aa.2005.107.3.377.

Filer, Colin, and Pierre-Yves Le Meur. 2017. *Large-Scale Mines and Local-Level Politics: Between New Caledonia and Papua New Guinea.* Canberra: ANU Press.

Gaillard, Catherine. 2010. *Développement minier et ruralité en Nouvelle-Calédonie.* Nouméa: IAC.

Gorohouna, Samuel. 2021. "Intégration de la population autochtone dans l'économie de marché autour de l'usine du Nord: une autre voie économique est-elle possible?" *Cahiers du LARJE*, No. 2020-4 15.

Gorohouna, Samuel, and Catherine Ris. 2015. "Inégalités ethniques sur le marché de l'emploi en Nouvelle-Calédonie." *Ethnies: Emancipations Kanak 37-38,* XXI, 122–131.

Gregory, Christopher. 1982. *Gifts and Commodities.* London: Academic Press.

Gregory, Christopher. 1997. *Savage Money: The Anthropology and Politics of Commodity Exchange.* Amesterdam: Harwood Academic Publishers.

Grochain, Sonia. 2013. *Les dynamiques sociétales de projet Koniambo.* Nouméa: Institut Agronomique Calédonien.

Guyard, Stéphane, Leïla Apithy, Severine Bouard, Jean-Michel Sourisseau, Michel Passouant, Pierre-Marie Bosc, Jean-François Belieres. 2013. *L'agriculture des tribus en Nouvelle-Calédonie. Résultats d'une enquête de 2010 sur la place et les fonctions de l'agriculture, l'élevage, la pêche et la chasse pour les groupes domestiques résidant en tribu.* Pouembout: IAC/CIRAD.

Hinojosa, Leonith. 2013. "Change in Rural Livelihoods in the Andes: Do Extractive Industries Make Any Difference?" *Community Development Journal* 48 (3): 421–436. doi: 10.1093/cdj/bst023.

ISEE. 2019. *General Census of Population.* New Caledonia, Nouméa: ISEE.

Kowasch, Matthias. 2010. "Les kanak face au développement de l'industrie du nickel en Nouvelle-Calédonie." PhD Thesis, Paul-Valéry Montpellier III - Heidelberg.

Le Meur, Pierre-Yves. 2021. "Industrie minière et fabrique voyageuse du développement." *Anthropologie & développement* Hors-Série 2021, 145–160. doi: 10.4000/anthropodev.1214.

Le Meur Pierre-Yves, Sonia Grochain, Matthias Kowasch, and David Poithily. 2012. *Sous-traitance minière et gouvernance locale en Nouvelle-Calédonie.* Rapport CNRT Nickel et son Environnement.

Levacher, Claire. 2016. "Penser la ressource minière en Nouvelle-Calédonie. Souveraineté, développement et valeur des lieux." *Développement durable et territoires. Économie, géographie, politique, droit, sociologie* 7 (3): 1–15. doi: 10.4000/developpementdurable.11429.

Lion, M. 2022. "Analyse des dynamiques spatiales des usines métallurgiques de Goro et Koniambo en Nouvelle-Calédonie." Master's Thesis, Institut d'Auvergne-Rhône-Alpes du Développement des Territoires - Université Clermont Auvergne.

Magrin, Géraud. 2013. *Voyage en Afrique rentière: une lecture géographique des trajectoires du développement.* Paris: La Sorbonne.

Mathieu, Laurent, Cécile Couharde, and Gilles Pestaña. 2016. "Cadrage macroéconomique et faits stylisés." In *La Nouvelle-Calédonie face à son destin,* edited by Séverine Bouard, Jean-Michel Sourisseau, Vincent Geronimi, Séverine Blaise, and Laïsa Ro'i, 69–109. Paris: Karthala.

Mazer, Katie, Suzanne Mills, and Séverine Bouard. 2022. "Pathways to Nickel Mining Employment among Inuit Women in Nunavik, Canada and Kanak Women in New Caledonia:

A Comparative Study." *The Extractive Industries and Society* 10, 101088. doi: 10.1016/j.exis.2022.101088.

Owen, John, and Deanna Kemp. 2017. *Extractive Relations. Countervailing Power and the Global Mining Industry*. Abingon: Routledge.

Pestaña, Gilles, Séverine Bouard, and Jean-Michel Sourisseau. 2013. *Ruralité et développement durable en Nouvelle-Calédonie: synthèse des ateliers de Poindimié et perspectives*. Païta (Nouvelle-Calédonie): IAC.

Roussineau, Élise. 2020. "Activité minière et niveau de vie: comment la proximité à la mine influence le niveau de vie local en Nouvelle-Calédonie?" Master 2 Thesis, Université de la Nouvelle-Calédonie.

Sabinot, Catherine, and Jean-Brice Herrenschmidt. 2020. "La dynamique des pratiques change-t-elle la manière de penser les relations kanak aux continuités terre-mer et à la nature?" *Anthropologie et Sociétés* 43 (3): 269–294. doi: 10.7202/1070157ar.

Sabinot, Catherine, and Shani Lacombe. 2015. "La pêche en tribu face à l'industrie minière dans le sud-est de la Nouvelle-Calédonie." *Revue de la Société de Biométrie Humaine* 5: 120–137.

Salaün, Marie. 2014. "Des Kanaks au Québec: l'expérience de jeunes autochtones calédoniens en formation aux métiers de la mine à Rouyn et à Sept-Îles." *Recherches Amérindiennes au Québec* 44 (2–3): 55–62.

Sourisseau, Jean-Michel, Cédric Gaillard, Séverine Bouard, and Valérie Angeon. 2021. "Mesurer les revenus agricoles en Nouvelle- Calédonie et en Guadeloupe. Méthodes et enseignements." *Notes et Études Socio-Économiques* 49 (Juillet 2021): 63–102.

Vadot, Guillaume, Christine Demmer, Séverine Bouard, and Mathilde Baritaud forthcoming. "Trajectories of Kanak Women in the Nickel Industry in New Caledonia." In *Mining Encounters and Indigenous Sustainable Livelihoods: Perspectives from the Americas, Fennoscandia and Oceania,* edited by Thierry Rodon, Alan Taylor, Sophie Theriault, Arn Keeling, and Séverine Bouard. Abington: Routledge.

Section IV

Conclusion

11 Mining and mobility

Key insights, governance implications, and future research

David Brereton, Gerardo Castillo Guzmán and Matthew Himley

Introduction

The contributions to this volume have explored mining-related mobility processes in diverse geographical settings, across different time frames, and from multiple theoretical perspectives. Collectively, the chapters have highlighted the various ways in which, over time, large-scale mining (LSM) *and* artisanal and small-scale mining (ASM) have shaped the spatial movements of people at different scales – sometimes driving these movements, sometimes enabling them, and at times constraining them. The chapters have also provided rich insights into how the mobilities put in motion by mining have changed both the character of places and the lives of people.[1]

In this final chapter, we draw out key findings and themes from the individual contributions, using the conceptual framework described in the next section to structure this discussion. The latter part of the chapter addresses implications for the governance of mobility behaviors and concludes with a discussion of priorities and opportunities for further research in this important and dynamic field of study.

Mining-related mobilities: an organizing framework

Mobility behaviors can be characterized in a wide variety of ways, including in terms of the *direction* of movements (into a place, away from a place, between places), the *duration/frequency* of movements (intermittent/episodic, seasonal, long-term/transitional), and/or the *range* of these movements (transnational, intra-country, local, vertical). Clearly, it would be impractical to explore all these dimensions in any detail in the space of one chapter. Instead, for the purposes of this overview, we have opted to focus on four aspects, as detailed below. The framework is not intended to be exhaustive but captures most of the mobility behaviors that are likely to be of interest to researchers – and practitioners – working in the mining space. It also aligns with the main themes explored in this volume.

1 *Inward flows.*[2] This refers to the movements of people *toward* mining areas, on either a long-term or temporary basis. Inward flows are a major focus of Banks and Schwörer's chapter on mining-induced in-migration in Papua New

DOI: 10.4324/9781003313236-15

Guinea (PNG) and Mususa and Peša's chapter on in-migration into the Zambian Copperbelt. Flows into ASM areas are addressed by Dessertine et al., in their chapter on artisanal and small-scale gold mining (ASGM) in Guinea and Côte d'Ivoire, and by Dunia Kabunga, Marijsse, and Geenen, in their chapter on small-scale gold miners in the Democratic Republic of the Congo (DRC). Most of the other chapters in the volume also make some references to inward mobility in its various forms.

2 *Circular and itinerant flows.* We use the term "circular flows" to describe the recurrent movement of people between two places (e.g., a mining locality and a home community) and "itinerant flows" to describe movements between multiple places (e.g., different ASM sites). Examples provided in this volume are as follows:

- "Migrant workers" who reside in the vicinity of a mine for long periods (often in company-provided camps, or barracks) and return to their home communities/countries only infrequently (e.g., over holiday periods and on annual leave breaks). While this form of mobility is no longer common, it was historically important in several of the countries discussed in this volume, including Zambia (Mususa and Peša), New Caledonia (Le Meur), Bolivia (Marston), and the DRC (Dunia Kabunga, Marijsse, and Geenen).
- "Self-managing commuters" who stay at or near a mine when they are working, and who regularly travel back to their home communities on non-workdays, using their own means. This form of mobility is a major focus of Bouard and Boudjema's chapter on New Caledonia (see also chapter by Le Meur) and is discussed in Damonte, Godfrid, and López's chapter on mining, infrastructure, and mobility in the Andes.
- "Seasonal miners," such as farmers from Guinea who periodically travel to ASGM areas to supplement their incomes, returning to their home communities between cycles (see especially the chapter by Dessertine et al.).
- "Itinerant miners," such as the gold miners from the DRC, as described by Dunia Kabunga, Marijsse, and Geenen. This pattern of behavior is similar to what Dessertine et al. characterize as "reticular mobility." (See also Jønsson and Bryceson 2009, on inter-site mobilities of artisanal gold miners in Tanzania, and Mkodzongi and Spiegel 2020, on artisanal mining in Zimbabwe.)

Another form of circular mobility in the mining industry is "long distance commuting." This is where workers live a considerable distance from the mine and are flown or driven to the site for a rostered length of time, generally staying in mine-provided accommodation while "on-site." Although not a major focus of this volume, this has become an increasingly common workforce management practice in LSM in the Global South, as well as in the Global North. Our chapter thus draws on the broader mining literature to include a brief discussion of it below.

3 *Local flows.* This aspect of mobility relates to how people move around and in local spaces, such as within the village or locality where they live, or between adjacent communities – what Damonte, Godfrid, and López refer to as "routine movements," or "short commuting" (in the case of workers). Local flows are

the primary focus of Castillo Guzmán's chapter on the La Granja project in Peru and are also addressed in some detail by Damonte, Godfrid, and López in relation to the Tintaya mine (Peru) and the División Andina project (Chile). Several other chapters include interesting granular data about local-level movements – for example, Le Meur's case study of the Thio region in New Caledonia, the "life histories" of Zambian Copperbelt residents presented by Mususa and Peša, and Dunia Kabunga, Marijsse, and Geenen's study of artisanal gold miners in the DRC.

4 *Outward flows.* Outward flows are often linked to the depletion of resources and the cessation, or winding down, of mining activity. However, people can move away at any point in the life of a mine or an ASM area. None of the chapters in this volume focus specifically on outward flows, but several touch on this aspect, mainly in discussions of how places have been impacted by downturns in mining activity (for example, Le Meur on New Caledonia; Dunia Kabunga, Marijsse, and Geenen on the DRC; Mususa and Peša on the Zambian Copperbelt; Castillo Guzmán on the community of La Granja in the northern Andes of Peru; and Marston on tin mining areas in Bolivia). Le Meur, Castillo Guzmán, and Damonte, Godfrid, and López also provide examples of mining-related displacement, and Bouard and Boudjema describe how some Kanak people left mining employment (and mining areas) because of concern over mining's environmental impacts.

In the following section, we highlight findings and insights from the chapters related to each of these four aspects of mobility. We then address two important overarching themes: complexity and agency.

Key learnings

Inward flows

As indicated, three chapters in the volume have a strong focus on inward flows into mining areas, including active and potential LSM sites in PNG (Banks and Schwörer) and on the Zambian Copperbelt (Mususa and Peša), and at ASGM sites in West Africa (Dessertine et al.). These are very different places, in terms of their geography, ethnic composition, culture, politics, history, and the types of mining undertaken, so care must be taken in drawing out generalizations or transferrable "lessons." However, there are common themes that can be identified.

The first is that many people discussed in the chapters who were drawn toward mining localities in PNG, Zambia, and West Africa came from agricultural areas where they faced challenging livelihood situations (e.g., static or declining incomes, limited services, pressure on land resources, exposure to climatic changes and fluctuations in commodity prices). To improve their circumstances, people were prepared to go to considerable lengths – and take substantial risks – in the hope of "getting ahead." These are similar, of course, to the factors that have contributed to high rates of urbanization across much of the Global South.

Second, the chapters indicate that it is not always the prospect of securing mining-related employment that attracts people to mining areas. In Zambia, as discussed by Mususa and Peša, many people who relocated to the Copperbelt used the skills and knowledge they had obtained from working in agriculture to provide goods and services to nearby mining towns, including by growing and selling food crops, gathering firewood for sale, making charcoal, and brewing beer. In PNG, according to Banks and Schwörer, people have been drawn to mining areas not just by the prospect (or hope) of finding work in mining-related fields but also by opportunities to set up local enterprises, such as selling betel nut or alcohol, and sometimes to leverage kinship and other social ties to claim a share of compensation payments and royalties.[3] Access to better services (schools, medical clinics, etc.) has also been a drawcard. For many of the artisanal gold miners from Guinea, studied by Dessertine et al., mining was seen as a way of supplementing incomes from farming, for a few weeks or months a year, rather than as an alternative "career."

Third, in each locality or region, physical *and* social connections have played an important role in creating what Banks and Schwörer characterize as "migration pathways" that facilitate inward flows. One example they provide is the Porgera gold mine in PNG's Western Highlands, where the key physical connector was the construction of the Highlands Highway in the 1980s, which made Porgera accessible to many people in the Western Highlands, and beyond. The key social factor, particularly in the early stages of mine development, was kinship ties, though over time in-migration expanded to include people with more tenuous connections to the Porgera Valley, creating what might be characterized as a "non-virtuous" circle (see below).

Mususa and Peša likewise point out that connections created through social ties and networks have played an important role in facilitating inflows of people from other parts of Zambia into the Copperbelt. Citing Schubert, Engel, and Macamo (2018), they also observe that investments in electrification, road, rail, telecommunications, and other nodal infrastructure (developed primarily for the mining sector) have facilitated greater connections between people and enabled them to move to and around the region. In the cases of Guinea and Côte d'Ivoire (see chapter by Dessertine et al.), family and community ties back to "source" villages have been mechanisms for disseminating information about opportunities (and risks) to prospective miners, recruiting people to work in the ASGM sector, and providing support – and a point of entry – for new arrivals. The proliferation of mobile phones (a physical connector) has made it easier for people to stay connected to home communities and for information to spread about new finds.

A final point: while physical and social connections may facilitate the movement of people to mining areas, they can have dysfunctional consequences for those already living there, as the Porgera example demonstrates. Banks and Schwörer describe how surrounding populations – and others living further away – have drawn on kinship ties and social obligations with the Ipili people, the traditional owners of the Porgera Valley, to gain access to the massive revenue flows from royalties and compensation payments and the new infrastructure being developed. Unfortunately for the Ipili, the arrival of these "relatives" in large numbers has made the

Ipili a minority, changed local power dynamics, and unsettled local identities and relationships.[4] This raises broader questions about the potential for the "agency" exercised by one group of actors to have negative consequences for others, a point we return to later.

Circular and itinerant flows

A key "take-out" from those chapters focused on circular and itinerant movements is that these forms of mobility have often provided benefits for people engaging in them (and, by extension, for their families as well).

In New Caledonia, Bouard and Boudjema observe that "local commuting" to the mining area has enabled Kanak mine workers to "keep a foot in each camp," often traveling regularly on weekends and between roster cycles to their traditional communities, located some two-to-three hours away. This has helped with maintaining connections to the land and with keeping one's place in the local social hierarchy. Bouard and Boudjema trace how these mobility practices have benefited the workers concerned (and their families) and have performed broader social functions, such as diversifying income sources, building connections with the wider world, and sustaining cultural practices. In a similar vein, Le Meur observes that greater mobility has led to increased circulation of money and products between urban and tribal areas in New Caledonia.

Another example of how circular mobility arrangements have been beneficial to participants (and those connected to them) is provided by Marston in her chapter on tin mining in the Norte Potosí region of Bolivia. She describes how in recent decades a significant number of Indigenous people from the region have found work in small-scale, cooperative mining. While the mines may be a considerable distance from miners' home communities, many travel home regularly to assist with planting and harvesting and to participate in local festivals, often bringing with them gifts of food and drink, supplies, and money.[5] This is an example of mobility functioning as a mechanism for distributing wealth from mining to other localities.

Circular mobility has also been a feature of ASGM in Africa, at least in some countries. For example, Dessertine et al. report that many of the people from rural communities in Guinea who travel to ASGM areas go there for only a few weeks or few months at a time and still consider their home village to be their main residence. For these people, working in ASGM is an opportunity to periodically supplement their income from farming, improve housing conditions, pay off a debt, finance an investment, or possibly fund a move to another, more desired, living location. Dessertine et al., citing Bolay (2017), observe that some miners also maintain dual residencies: one in the mining locality, the other in their village of origin.

For artisanal gold miners adopting an itinerant lifestyle, the ability to move with relative ease from site to site has helped people to earn a living from mining over the longer term – including in contexts with limited alternative options (see chapter by Dunia Kabunga, Marijsse, and Geenen; also, Jønsson and Bryceson 2009). However, this remains a risky pathway, and rates of failure can be high (see, for example, Bryceson, Jønsson, and Shand 2020; Mkodzongi and Spiegel 2020).[6]

It is not just artisanal miners and mine workers who engage in circular mobility practices but also other people who reside in or near mining areas. For example, Castillo Guzmán, in his chapter on La Granja in Peru, reports being told by a senior officer of the company that by 2022 around 80 percent of La Granja's families maintained a second residency, primarily in the city of Chiclayo where their children were being educated. Mining-related income, mainly in the form of wages earned working for the company or associated sub-contractors, or from rents in some cases, had helped in establishing and maintaining these arrangements. (Having a base in Chiclayo may also have been a form of insurance in the event that the project did not proceed.)

The case of long-distance commuting

None of the chapters in the volume pay specific attention to long-distance commuting, which is also a form of circular movement. However, some observations are in order, as this form of workforce management is increasingly utilized by large mines world-wide.

Attracting the workforce needed to build and operate mines, and to provide necessary support services, has often been a challenge for the LSM sector. This has particularly been so for mines distant from large population centers and with climatic and geographical disadvantages (excessive heat or cold, high altitude, lack of water, exposure to diseases, etc.). Often in cases like these, the labor-supply issue was initially addressed through the use of forced migration, as in several areas studied in this volume, such as Bolivia (Marston), the DRC (Dunia Kabunga, Marijsse, and Geenen), and New Caledonia (Le Meur). Another common practice was to import foreign workers on fixed-term contracts, often encouraging white European migrants because they were considered more reliable, productive, etc. (see Le Meur's chapter). These workers were generally required to stay with the same employer for the duration of their contract.

In the twentieth century, the focus shifted toward creating incentives for people to come and work in these areas. This was often done by building dedicated mining towns that offered relatively good services and living conditions and enabled families to stay together. Town and infrastructure construction were actively encouraged – sometimes required – by governments, who saw this as a means of promoting development in remote areas (see chapter by Damonte, Godfrid, and López on the "lure of modernity"). Typically, constructing closed towns also served to insulate – and "protect" – workers and their families from local populations, and to maintain racial separation, as well as being used to build a company identity.[7]

The residential model has now been largely supplanted by long-distance-commuting arrangements in which workers are flown in – or bussed to – the mine from other locations, such as a capital city or regional center (or from other countries), and are accommodated on or near the mine site. Sometimes, this takes the form of a hybrid model in which most specialists and skilled workers commute, and a substantial proportion of the less-skilled jobs are filled by local people. In these cases, employment of locals is often driven more by social and political imperatives than by efficiency considerations.

Key drivers and enablers of the shift toward long-distance commuting include the rising cost of constructing and maintaining towns, changed workforce preferences, company efforts to weaken labor unions (often also associated with the increased use of contractors), and improved transport infrastructure that has facilitated the movement of more people over longer distances. Another justification sometimes advanced by companies is that long-distance commuting arrangements, using on-site accommodation facilities, place less stress on local communities and mitigate the risk of uncontrolled population inflows. More generally, this shift can be seen as part of a broader transition away from "nationalist" resource development policies and toward more free-market approaches, as suggested by Damonte, Godfrid, and López in their chapter.

Long-distance commuting is well established in countries as diverse as Australia, Canada, PNG, Chile, Peru, and Mongolia. However, it has not been universally accepted. It increases the flow of people between the mine and other – often distant – places but reduces local interactions and connectedness with the place where mining occurs. The ability of mine workers to live in a larger center can provide greater economic opportunities for workers' partners (mainly women) and give them greater freedom of movement, but comes at the cost of the mine worker being absent from the home for longer periods. There is also less potential for communities close to mines to benefit economically under this model, as wages and salaries are now mostly spent in and around the cities and towns that act as commuting nodes.[8]

While there is now a substantial literature on the social and economic impacts of long-distance commuting in Australia and Canada (e.g., Haslam-McKenzie 2016), the findings are not necessarily transferrable to other contexts. This presents an opportunity for mobility researchers working in other parts of the world, such as Latin America and Asia, to explore in more detail how this way of organizing work impacts "host" and "source" communities.

Local movements

Several chapters address how people exercise mobility *within* specific localities and regions. For instance, Damonte, Godfrid, and López, in their study of the Tintaya-Antapaccay mine in Peru, discuss how the mining company, with the support of the state, initially adopted an exclusionary approach to managing the area around the mine that included forced relocations. Over time, however, accommodation infrastructure developments and changes in company policy encouraged people from villages further away to move to the area and set up a second residence in the nearby town of Yauri. This gave them access to better services and improved their chances of finding mining work and securing a better house.

In another example, Castillo Guzmán documents how the La Granja project, even at an early stage in its life cycle, brought about significant changes in the local physical and social landscape. He argues that these changes, in turn, had implications for local women and the degree of autonomy they were able to exercise. While still disadvantaged relative to men, women reported being able to move more freely around public spaces than previously, in part because lighting and facilities had been improved and also because there were more people around. There were also fewer violent fights among men in the public space and less excessive drinking, partially due to corporate labor measures and restrictions. In addition, Castillo Guzmán reports that those women who found work at the project or in the businesses that serviced it were able to exercise more personal autonomy, including by being able to fund their own travel to places like the coastal city of Chiclayo, the capital city of Lima, and even internationally.

These studies provide more nuanced pictures of how people experience mining development than do most large-scale impact analyses. Castillo Guzmán's chapter is also a timely reminder of the importance of applying a "gender lens" to the analysis of mobility behaviors.

Outward flows

People can – and do – move away from mining areas at any point in the mining life cycle. For example, when a mine is being developed, or an ASM area is opening up, some people may be displaced from their traditional lands; others may leave in response to rising prices and rents, declining environmental conditions, and/or because they do not like the social changes occurring. Some area residents may use compensation payments, or income from land sales or rents, to fund a move to another, more preferred locality. During a mine's operational phase, people may move away for varied reasons, including because they lose their job due to workplace re-structuring, to improve their economic circumstances, to access better amenities and services, or for family-related reasons. Toward the end of the development cycle, when mines can no longer operate profitably, or easily accessed alluvial resources are exhausted, there will often – although not always – be population outflows as people seek opportunities elsewhere.

Although some chapters provide historical examples of closures and industry downturns contributing to population outflows, the volume mainly focuses on active mining areas. Hence, it is understandable that more attention has been directed to inflows and circular flows than to outward flows. There are also practical obstacles to studying outward flows, particularly in the ASM sector (as discussed by Bryceson, Jønsson, and Shand 2020). When people move away from an area, they typically disperse and can be much harder to locate – *places* are generally easier to study than are *people on the move*. As we discuss below, in the future "tracking" studies could provide valuable data about where people go to – and what they do – when they move away from mining areas, though these studies will likely be expensive and time consuming.

When studying the impacts of mine closure, it is important to focus on not just who moves but also who does not (see Marais 2023). Do people who stay lack the capacity to move, or do they do so out of preference? If it is the former, what is impeding their mobility, and what can be done to address this? How do non-movements impact places? For example, do they contribute to the entrenchment of poverty? Given that a large number of mines around the world are likely to cease production in upcoming decades – due to either ore depletion or closure time-frames imposed by governments (for coal mines in particular) – these questions will become increasingly salient topics for mobility researchers.

Overarching themes

Our focus to this point has been on how different aspects of mobility have been addressed in the book. In this section, we highlight two important overarching themes that have emerged.

The first is that, when viewed as a whole, mining-related mobility processes are extremely *complex*. They involve a lot of people; they are shaped by a multiplicity of factors, not all of which are linked to mining (see especially Mususa and Peša's chapter); and they function in different ways depending on local context. This all might seem obvious, but it bears repeating given the propensity of some researchers to seek explanations with general application, and of some practitioners and policy makers to assume that governance strategies effective in one context will work equally well in others.

To take this a step further, it could be argued that some issues arising in relation to mobility behaviors (e.g., managing large-scale population inflows into mining localities; improving working conditions and reducing environmental damage in ASM areas) are so complex that they constitute "wicked problems." This concept, introduced into academic discourse by two University of California planning professors, Rittel and Webber (1973), proposes that some problems are not amenable to solutions in any absolute sense: "solutions" are only better or worse. What makes a problem wicked is that its causality is entangled, it is often a manifestation of other problems that have their own causal structures, different actors have different understandings of the problem and how it might be addressed (and may disagree on whether there is a problem at all), and "solutions" cannot be assumed to be transferrable from one context to another (and will likely have uneven effects on actors involved). Furthermore, any action taken to address a wicked problem is highly likely to impact future behavior, whether intended or not – thus, there is no opportunity to experiment "without regrets."

Framing an issue as a "wicked problem" is never an excuse for inaction, but it does serve as a reminder of: (a) the importance of investing time and effort into understanding the underlying causal complexity of the issue being addressed; and (b) the need to accept that some problems cannot be solved in an absolute sense and that any "solutions" will be sub-optimal (with uneven effects for those involved) and, most likely, timebound. Moreover, "wicked problems" typically cut across organizational boundaries and therefore collaboration is generally required to better understand "the problem" and to devise and implement possible responses. (These points are highly relevant to the next section, "Mobility and governance.")

The second theme to highlight is the importance of *human agency*, or what Bainton (2017, 315) characterizes as "dialectical resourcefulness." Commonly, when people decide to move – individually or collectively – they are not responding passively to changes in their broader environment, but rather strategizing, making trade-offs, and consciously taking risks. In other words, they are exercising agency. Examples from chapters in the volume include the following:

- PNG villagers moving to areas, even before mine-development is confirmed, to position themselves to take advantage of possible future opportunities (Banks and Schwörer).
- Zambian farmers relocating to the Copperbelt and seeing opportunities to establish small businesses to provision nearby mining communities (Mususa and Peša).
- Local landowners who had moved away from the area around the Tintaya mine in Peru returning to take part in negotiations for compensation and land access (Damonte, Godfrid, and López).
- Bolivian tin miners strategically engaging in capitalist economic activities to support contemporary expressions of Indigenous autonomy (Marston).
- Peruvian women using the opportunities provided by the La Granja project to exercise greater autonomy (Castillo Guzmán).
- Kanak people in New Caledonia establishing informal dual residency arrangements to spend time in their traditional communities while continuing to work in the mining industry (Bouard and Boudjema).

Attention to human agency in situations like these is especially important given the economic, social, and political contexts in which these actors are operating – i.e., in situations of limited economic opportunities, uneven governmental support, and/or long histories of socio-economic marginalization. In contexts like these, individuals and families are often heavily reliant on their own resources and initiative to sustain livelihoods and, ideally, improve their position over time. This is not to suggest, of course, that every act of agency is "rational" in the conventional sense of the term – only that people strategize and make choices.[9]

We must also be cognizant that the consequences of one group of people exercising agency can sometimes be detrimental to the rights and well-being of other groups. The proposed Wafi-Golpu mine in PNG offers an example: the creativity of other Papua New Guineans in finding ways to get a foothold in the area has arguably been at the long-term expense of the traditional landowners. Another example concerns people from Burkina Faso who have moved to Côte d'Ivoire and established small-scale gold extraction businesses (see chapter by Dessertine et al.). Again, they have demonstrated considerable agency, but their businesses rely heavily on child labor, mostly brought in from villages in Burkina Faso. It is also important not to overestimate the leverage of agency when people – and women in particular – face structural imbalances and disadvantages (see chapter by Castillo Guzmán).

Mobility and governance

As several of the chapters in this volume have demonstrated, mobility can provide people with a means of managing risks and building resilience. This has also been an important theme in studies of risk aversion and diversification among peasant populations (for example, Roumasset 1976; Scott 1976).

For Guinean farmers, traveling to different areas to engage in ASGM on a "seasonal" basis has become an established strategy for diversifying income sources and accumulating additional capital (see Dessertine et al. in this volume). In New Caledonia, the ability of Kanak mineworkers to move between their traditional territories and the mining areas has reduced the vulnerability of their communities to fluctuations in incomes from traditional activities, such as farming or fishing, while also allowing the workers to maintain alternative income sources in the event of mine closure or job loss (see chapters by Le Meur and Bouard and Boudjema). Similarly, Marston observes that droughts, combined with difficult economic conditions at the national level in Bolivia, have led Indigenous communities to look to mining as an economic activity with stabilizing potential, which has entailed community members traveling to mining areas in other localities for work. At a strategic level, this has also sparked efforts by Indigenous people in Bolivia to frame, or re-frame, subsoil resources as part of Indigenous territory.

From a "system" perspective, mobility may thus be viewed as a mechanism for re-allocating labor away from places with few jobs and stressed local production systems to other areas where the demand for labor is greater and there are more opportunities. This mechanism can also potentially operate in reverse when mines close or alluvial resources are depleted: people move away to other localities or into other sectors where there are better opportunities, rather than being stranded in a place in decline. (Economists refer to this as "market clearing," although real world labor markets are often much stickier than economic theory assumes.)

Another potentially beneficial aspect of mobility is that it broadens connections between mining areas and other localities and, in doing so, can contribute to the breakdown – or at least weakening – of the traditional mining "enclave." Inward movements can bring new skills and capacities into a region, as the chapter by Mususa and Peša shows in the context of the Zambian Copperbelt (for a Chilean example, see Rodrigo, Romani, and Ricci 2018). Circular movements likewise diffuse income generated from mining across wider areas and create broader linkages. An example of the latter is provided by Le Meur's chapter, which argues that the movement of people between mining areas, and between these areas and the capital Noumea – combined with changes in the political and administrative apparatus of the country – has reconfigured New Caledonia as a single enclave (see also chapter by Bouard and Boudjema).

From a policy perspective, then, there will be circumstances where governments and companies should be seeking to facilitate mobility, rather than restricting it: such as by removing physical and legal restrictions on mobility, providing people with skills and qualifications that can be utilized in other settings, and reducing place-based dependencies. Marais, Cloete, and Lenka (2022) have advocated for this approach in relation to mine closure in South Africa, as a means of reducing the risk of creating poverty traps in mining-dependent areas. This approach may also be relevant to regions such as Appalachia in the United States, which is already experiencing large-scale closures as the transition away from coal as an energy source gathers pace.

Mobility behaviors become problematic when they impact adversely on the rights of others and cause or exacerbate unwanted social and environmental changes. This can occur when many people make what might be individually rational decisions to move to a locality, but local socioecologies are overwhelmed by the scale and speed of the influx. The LSM sector provides several examples of the adverse impacts that can result, including urban over-crowding, rising house prices, growth in informal settlements, environmental degradation, increased social tensions, higher crime levels, and the weakening of traditional cultures and authority structures (see chapter by Banks and Schwörer; also, IFC 2009; Bainton et al. 2017; Wilson 2022).

In the ASM sector, the phenomena of "rushes" or "booms" can likewise be disruptive for local communities, especially in the short term, although this may ease as inflows slow and there is some stabilization of relations between miners and local communities. Dessertine et al.'s chapter describes how the inflow of tens of thousands of people within a few days into a "new" mining region in northwestern Guinea set off a train of events resulting in a climate of violence between gold miners, local people, and the military. However, the situation improved after management of the mines was transferred from the military to the local authorities and accommodations were reached with newly arrived miners (see also Damonte 2021 on Peru's Madre de Dios region).

Contextual factors may add to the complexity of large-scale population influx dynamics and may impact the options available to address associated problems. Many large-scale population inflows into mining areas (both LSM and ASM) seen in recent years have occurred in relatively poor countries with limited economic opportunities for much of the population, especially those living outside of major population centers.[10] Many of these countries have histories of uneven development and colonialism, and of foreign exploitation of mineral resources. Moreover, rushes have frequently occurred in areas where state agencies have limited reach and effectiveness, lack the resources, capabilities, and, in some cases, motivation, to establish and maintain a stable governance environment, and where local political processes and power structures can be unstable – what Boege, Brown, and Clements (2009) characterize as hybrid political orders.[11]

We now briefly consider what might feasibly be done to improve management of the scale and speed of population influxes into mining areas and mitigate adverse impacts. Our primary focus is the LSM sector, but we also make brief observations about the ASM sector.

Improving influx management in the LSM sector

In 2009, in response to growing concerns about the impacts of project-induced in-migration (PIIM) in the LSM sector – and also the business and political risks it presented for companies and governments – the International Finance Corporation (IFC) produced a *Handbook for Addressing Project-Induced In-Migration* (2009).

In addition to providing guidance for assessing the likelihood of PIIM causing adverse social, health, and environmental impacts, the *Handbook* identifies strategies that could be used to contain and reduce inflows, including moving existing settlements further away from the mine, restricting road access, operating mines on a fly-in/fly-out basis, promoting diversified regional growth strategies (including alternative growth centers located away from the mine), and assisting immigrants to relocate to other areas.[12]

While the advice provided by the IFC is generally regarded as sound by industry specialists (Bainton et al. 2017, 4), uptake in the mining sector appears to have been patchy. According to Bainton et al. (2017, 6), "The general knowledge base on the management of PIIM remains shallow and under-developed," and "PIIM management plans are rarely publicly available." Although the state has a potentially crucial role to play in this area, there was apparently no equivalent guidance produced for governments on: (a) how to independently assess PIMM risks as part of the project approval process; or (b) measures that governments themselves can take to help manage these risks.

In their chapter on PNG, Banks and Schwörer conclude that corporate and state efforts to control population inflows into places like Porgera, and to a lesser extent at Lihir, have not been successful; nor have attempts by mining companies to mitigate the impacts of these flows. In part, this is due to internal management failings in companies, as suggested by Banks and Schwörer (see also Owen and Kemp 2017), but large-scale population influxes also have structural characteristics which make the adverse effects inherently difficult problems to manage, let alone solve (see preceding discussion on "wicked problems"). Companies can exercise some control over the factors that draw people to mining sites, but they typically have limited influence over the broader socio-economic forces "pushing" people to move away from traditional lifestyles and toward mining areas. Furthermore, as already noted, the people whose behaviors companies may try to control are generally motivated to seek out opportunities and have demonstrated a high capacity for ingenuity, risk taking, and persistence.

Improving the social performance management capabilities of companies and increasing the amount and quality of information available to decision-makers will help. However, once people begin moving into areas in large numbers, it can be very difficult to reduce, let alone reverse, these flows. Instead, there needs to be much greater scrutiny of project proposals *before* they are approved, particularly proposals to locate mines in places where there is an appreciable risk of large-scale population inflows if a mine is constructed. In some instances, the risks of proceeding may be high enough to make proposal rejection the only responsible course of action; in other instances, there may need to be a major reconfiguration of the mine itself (e.g., reducing the scale of the project, approving an underground operation only, or relocating key infrastructure).

Governments alone cannot be relied on to provide this level of oversight, as they are often a conflicted party (especially in countries with high levels of economic and fiscal dependence on mining). If progress is to be made, it will require the involvement of a broader array of actors – including informed community members, civil society organizations, experienced social performance practitioners, financial risk analysts, and academic researchers – in "shining the light" on project proposals and highlighting risks they present. As befits a "wicked problem," this is a "clumsy" and partial solution, but it is hard to see viable alternatives.

Where a proposed project is located on or close to the lands of Indigenous peoples, it should also be a requirement to obtain the free, prior, and informed consent of these groups before proceeding. As part of this process, communities must be informed about the likelihood of a significant influx of people into the area if the project is approved, and what the environmental and social consequences might be. This information must be communicated using words, concepts, and ways of presenting data that people understand.

The ASM sector

Limiting and regulating population inflows is arguably even more challenging with ASM, as demonstrated by the mostly failed attempts of governments to formalize the sector (Hilson et al. 2017). Also, when states seek to directly regulate ASM, violence often results. ASM tends to be highly decentralized – the key actors are typically loosely connected groups of miners, not large companies – and mining is often undertaken in areas of uneven or ineffective state control. Consequently, standard mechanisms for controlling and moderating the behavior of miners – such as leases, licenses, and planning rules – cannot be deployed. (For recent contributions, see Damonte 2021 and Giraldo Malca et al. 2023, on the inability of the Peruvian state to effectively regulate and "formalize" ASGM in the Peruvian Amazon, and Kalokoh and Kochtcheeva 2022, on Liberia and Sierra Leone.)[13] Another significant ASM governance issue, which Mkodzongi and Spiegel (2020) highlight in the case of Zimbabwe, is that miners can be vulnerable to abuse and exploitation by criminal syndicates and powerful political and business interests.

We acknowledge that a weak or absent state should not necessarily be equated with a lack of governance. In some settings, local traditional authorities and other actors govern ASM activities by reaching accommodations with miners that mitigate the impacts of population inflows and ensure that local landowners derive economic benefits (see Dessertine et al.'s discussion of the role of *tombolomaw* in ASGM areas in Guinea and Damonte 2021 on Peru's Madre de Dios region). These arrangements are context specific, may be hard to replicate, and often require trade-offs. But given the scale of the ASM sector and its importance as a source of income for millions of people, it is essential to continue exploring the potential to adapt – or create – local-level governance structures and processes to better manage population flows into ASM areas and the impacts of these, and to protect miners themselves from violence and exploitation. Based on the evidence to date, strengthening local institutions and norms is likely to be a more feasible way forward than initiatives that rely on the "top down" exercise of state power – or, for that matter, than efforts to drive improved practices through mineral certification schemes. A key first step is to understand the often-complex ways ASGM activities and mobilities are governed in practice at the local level, and to ascertain what might be achievable in particular contexts.

Directions for future research

As this volume has demonstrated, the study of mining-related mobilities, in both the ASM and LSM sectors, is a vibrant and important area of research. Here, we make four suggestions for how to extend – and in some cases refocus – this agenda.

1 *Set up baseline and tracking studies in mining areas to monitor mobility dynamics – and associated social changes – over time.* Establishing and maintaining the appropriate data collection processes will require substantial resources, but some research funding bodies (e.g., the Ford Foundation) and international agencies (e.g., the World Bank) may be open to supporting this kind of work. It may also be possible to interest some governments – and even some mining companies – in enabling this work, on the basis that it is in their long-term interests to better understand the social environments of mining areas. Engaging with the broader academic research community will likewise be important. Longitudinal studies are now relatively commonplace in health research and some areas of sociology; there is a strong case for more thoroughly extending this approach to the mining domain as well.[14]
2 *Focus on people who move, as well as the places that they move to or from.* Adding a perspective in which families and social networks, rather than physical places, are the primary units of analysis, would provide a broader understanding of the complex relationship between mining, mobility, and social change (see Castillo and Brereton 2018, 469). Again, setting up such studies would be challenging, particularly in the ASM sector, but it should be possible – at least in principle – to establish long-term "panel studies," where the same cohort of study participants is periodically re-interviewed. This would provide a way to stay in touch with people who have moved into or out of mining areas, in order to understand how their lives – and those of others – are affected in the longer term. Again, these kinds of studies have been undertaken successfully in other research domains, such as public health.
3 *Focus on those places that are connected to mining areas by mobility, as well as on the mining areas themselves.* As we have documented in this overview, mobility often involves people moving away from their home communities to mining areas and, in the case of circular mobility, between mining and non-mining spaces. This can lead to significant social and economic changes in the source communities. Economically, they can benefit from the wealth generated by mining work, including through remittances, gifts, investments, and the diversification of income sources. On the other hand, mobility may, in some instances, be a mechanism for transmitting harm between localities (for example, by contributing to the spread of diseases in source communities and, in the case of ASM, increasing the use of child labor). Mining-related mobility can also change the demographic structure of source and destination localities, for instance by

drawing men away from their home areas and, conversely, contributing to the masculinization of destination localities. The role that mining-related mobility processes play in driving and enabling social change across geographies should be a productive area for further research.

4 *Study immobility, as well as mobility.* In order to understand what makes people move, we also need to understand why people choose not to move or are prevented from moving. As we have noted, this is emerging as a significant issue for localities and regions that are experiencing, or will soon experience, the effects of mine closures. People may decide to stay in an area that has been – or will be – impacted by closure because they have other income sources, have strong social connections, or are simply attracted to the local area. Others may feel trapped because they cannot envision alternatives, do not possess transferrable skills, or lack the confidence or networks to make a new start elsewhere. Understanding these barriers – and how to deal with them – will be essential if the aspirational goal of "just transitions" is to have practical application. Immobility can also be a major issue in active mining areas: for example, where women are restricted in their public movements, or trapped in unequal and abusive relationships.

Conclusion

When people move around, things change. Populations can increase or decrease, sometimes drastically; the demographic composition and socio-political dynamics of areas can be altered; conflict within and between groups may increase, or people may find ways to live together; economic opportunities may be expanded or diminished; ideas and technologies can spread, and traditional knowledge can be lost; people can have their rights violated in some settings and be able to exercise greater freedom in others. Expected changes in the mining landscape in upcoming decades will add further complexity to this picture.

On the one hand, there will be an increasing number of mine closures – not just in the coal sector but also in metalliferous mining, where a growing number of mines are coming to the end of their operating lives. This will likely lead to increased population outflows from some mining areas, although, in some cases, the problem may prove to be not too much movement, but too little. On the other hand, demand for "critical and strategic minerals" – e.g., copper, cobalt, nickel, and lithium – is predicted to grow dramatically, as the world transitions to a post-carbon economy (International Energy Agency 2022). Meeting this demand will require the construction of new mines, including in areas which are economically underdeveloped, are already experiencing pressure on land and resources, and have a history of social and political instability, and will likely lead to greater encroachments on the lands of Indigenous people (Lèbre et al. 2020; Extractive Industries Transparency Initiative 2022). Growing demand could also lead to an increase in ASM in some regions, especially for minerals like cobalt and tantalum. There are also predictions of a resurgence of mining in rural areas in Europe (Del Mármol and Vaccaro 2020).

Overall this volume has showcased an impressive body of research on the mining-mobility nexus, covering three major mining regions of the world, and both the ASM and LSM sectors. As the industry continues to evolve and change, there will be ample opportunities – and growing need – for researchers to build on this foundation, to explore new questions, and to use this knowledge to help improve outcomes for people and places that have been – or are likely to be – impacted by mining.

Notes

1 Although our focus in this chapter is on human mobility, we note that the term has also been used to describe how other elements "travel" across the landscape, including technologies, money, waste, and pollutants, and even minerals themselves. (See chapters by Dunia Kabunga, Marijsse, and Geenen on gold as a mobile element and Le Meur on the movement of mining contaminants across the landscape in New Caledonia.)

2 We prefer the term "inward flows" (or "inflows") over "in-migration." The latter implies relocation, whereas the concept of "inflows" includes temporary stays.

3 Banks and Schwörer also refer to two proposed mining projects in PNG – Wafi-Golpu and Frieda River – where people have moved into adjoining areas even before confirmation that the projects are going ahead, with the aim of positioning themselves to access jobs, compensation and royalty payments, and business opportunities, should the mine proceed.

4 The inverse of social connectors are social barriers, which can inhibit the movement of people into new areas. For example, Marston, in her chapter, describes how Indigenous miners in Bolivia had to overcome ethnic divisions and racism to mine under the legal auspices of mining cooperatives.

5 This description has some parallels with the historically significant figure of the peasant-miner in the central Andes of Peru (see DeWind 1987; Long and Roberts 1984).

6 Bryceson, Jønsson, and Shand (2020, 460) note that many of the Tanzanian gold miners they interviewed had improved their material circumstances but add the important rider: "There are uncounted others, not on hand to be interviewed, who have crashed and returned to their villages or went elsewhere, abandoning mining with a sense of loss and wasted effort."

7 In their chapter, Dunia Kabunga, Marijsse, and Geenen provide the example of a town in the DRC, where a group of young men called themselves *les garcons de Kaga* (the boys from Kamituga) and identified themselves as "sons" of the company.

8 See, for example, recently published research by Prada-Trigo, Barra-Vieira, and Aravena-Solís (2021) who attribute the strengthening of the real estate market in the Concepción metropolitan area in Chile to the presence of a large number of commuting mine workers, many of whom were in the process of buying houses.

9 For example, Bryceson Jønsson, and Shand (2020) describe the behavior of artisanal gold miners in rushes as sometimes akin to "irrational exuberance," where prospects of success are over-stated, and risks ignored or downplayed.

10 As documented in the literature on "boomtowns," some resource-rich areas in wealthier countries have also experienced rapid and difficult to manage population inflows, particularly in the hydrocarbons sector (Jones and Mayzer 2021). However, the impacts, while disruptive, have not been on the same scale as in places like PNG or West Africa.

11 According to Boege, Brown, and Clements (2009), a hybrid political order, which is most likely to arise in historical formations where the state is fragile, is characterized by a combination of traditional social structures, elements of the Western model of the state, and current organizations and movements originated by and reacting to globalizing forces.

12 On one view, the IFC's approach could be interpreted as an attempt to return to an "enclave model," by trying to limit access to outsiders. Another option could be to take an integrated regional planning approach and work to create a more diverse local economy, but this would probably present formidable governance challenges. It is also the case that mines are often located in areas that otherwise would be unlikely to attract large-scale investment.

13 There are also examples of governments changing laws and policies in ways that, intentionally or otherwise, reduce rather than strengthened controls over movements into mining areas. As Dunia Kabunga, Marijsse, and Geenen discuss, in the DRC in the 1970s-80s, in the context of economic and political collapse, mining laws were liberalized, and Congolese people effectively encouraged to "fend for themselves," including by engaging in ASM.

14 A potential model is the demographic monitoring program set up by Lihir Gold in PNG in the 1990s. This program collected detailed longitudinal information on the number of Lihirians and non-Lihirians on the island and so could be used to track the extent of in-migration, including whether the migrants were "returning" Lihirians or from elsewhere in PNG. The database also recorded health and other socio-economic data (see Bainton 2017).

References

Bainton, Nicholas. 2017. "Migrants, Labourers and Landowners at the Lihir Gold Mine, Papua New Guinea." In *Large-Scale Mines and Local-Level Politics: Between New Caledonia and Papua New Guinea*, edited by Colin Filer and Pierre-Yves Le Meur, 313–351. Canberra: Australian National University Press.

Bainton, Nicholas, Vlado Vivoda, Deanna Kemp, John Owen, and Julia Keenan. 2017. *Project-Induced In-Migration and Large-Scale Mining: A Scoping Study*. Brisbane: Centre for Social Responsibility in Mining, The University of Queensland: https://www.csrm.uq.edu.au/publications/project-induced-in-migration-and-large-scale-mining

Boege, Volker, Anne Brown, and Kevin Clements. 2009. "Hybrid Political Orders, Not Fragile States." *Peace Review* 21 (1): 13–21. doi: 10.1080/10402650802689997.

Bolay, Matthieu. 2017. "Gold Journeys: Expulsion-Induced Mobility and the Making of Artisanal Mining Spaces in West Africa. An ethnography of Itinerant Labour at the Bottom of the Gold Supply Chain.' PhD diss., Neuchâtel University.

Bryceson, Deborah Fahy, Jesper Jønsson, and Mike Shand. 2020. "Mining Mobility and Settlement During an East African Gold Boom: Seeking Fortune and Accommodating Fate." *Mobilities* 15 (3): 446–463. doi: 10.1080/17450101.2020.1723879.

Castillo, Gerardo, and David Brereton. 2018. "Large-Scale Mining, Spatial Mobility, Place-Making and Development in the Peruvian Andes." *Sustainable Development* 26 (5): 461–470. doi: 10.1002/sd.1891.

Damonte, Gerardo 2021. "Limited State Governance and Institutional Hybridization in Alluvial ASM in Peru." *Resources Policy* 72: 102118. doi: 10.1016/j.resourpol.2021.102118.

Del Mármol, Camila, and Ismael Vaccaro. 2020. "New Extractivism in European Rural Areas: How Twentieth First Century Mining Returned to Disturb the Rural Transition." *Geoforum* 116: 42–49. doi: 10.1016/j.geoforum.2020.07.012.

DeWind, Josh. 1987. *Peasants Become Miners: The Evolution of Industrial Mining Systems in Peru*. New York: Garland Publishing.

Extractive Industries Transparency Initiative. 2022. *Making the Grade: Strengthening Governance of Critical Minerals*. Oslo: Extractive Industries Transparency Initiative.

Giraldo Malca, Ulisses Francisco, Ana Sabogal Dunin-Borkowski, Nicolas Facho Bustamante, María José Mori Reaño, and José Miguel Giraldo Armas. 2023. "Alluvial Gold Mining, Conflicts, and State Intervention in Peru's Southern Amazonia." *The Extractive Industries and Society* 13: 101219. doi: 10.1016/j.exis.2023.101219.

Haslam-McKenzie, Fiona. 2016. "The Socio-Economic Impacts of Long-Distance Commuting on People and Communities." In *Labour Force Mobility in the Australian Resources Industry Socio-Economic and Regional Impacts*, edited by Fiona Haslam-McKenzie, 11–28. Singapore: Springer.

Hilson, Gavin, Abigail Hilson, Roy Maconachie, James McQuilken, and Halima Goumandakoye. 2017. "Artisanal and Small-Scale Mining (ASM) in Sub-Saharan Africa: Re-Conceptualizing Formalization and 'Illegal' Activity." *Geoforum* 83: 80–90. doi: 10.1016/j.geoforum.2017.05.004.

International Energy Agency. 2022. *The Role of Critical World Energy Outlook Special Report Minerals in Clean Energy Transitions*. Paris: International Energy Agency.

International Finance Corporation. 2009. *Projects and People: A Handbook for Addressing Project-Inducted In-Migration*. Washington, DC: International Finance Corporation.

Jones, F. Matt, and Roni Mayzer. 2021. "A Fifth Wave? The Changing Nature of Boomtown Research." *The Extractive Industries and Society* 8 (2): 100893. doi: 10.1016/j.exis.2021.02.013.

Jønsson, Jesper, and Deborah Fahy Bryceson. 2009. "Rushing for Gold: Mobility and Small-Scale Mining in East Africa." *Development and Change* 40 (2): 249–279. doi: 10.1111/j.1467-7660.2009.01514.x.

Kalokoh, Amidu, and Lada V. Kochtcheeva. 2022. "Governing the Artisanal Gold Mining Sector in the Mano River Union: A Comparative Study of Liberia and Sierra Leone." *Journal of International Development* 34 (7): 1398–1413. doi: 10.1002/jid.3643.

Lèbre, Éléonore, Martin Stringer, Kamila Svobodova, John R. Owen, Deanna Kemp, Claire Côte, Andrea Arratia-Solar, and Rick K. Valenta. 2020. "The Social and Environmental Complexities of Extracting Energy Transition Metals." *Nature Communications* 11 (1): 4823–4823. doi: 10.1038/s41467-020-18661-9.

Long, Norman, and Bryan Roberts. 1984. *Miners, Peasants and Entrepreneurs: Regional Development in the Central Highlands of Peru*. Cambridge University: Cambridge.

Marais, Lochner. 2023. *The Social Impacts of Mine Closure in South Africa: Housing Policy and Place Attachment*. Abingdon: Routledge.

Marais, Lochner, Jan Cloete, and Molefi Lenka. 2022. "The Plight of Mining Cities in South Africa: Planning for Growth and Closure." *Cities* 130: 103965. doi: 10.1016/j.cities.2022.103965.

Mkodzongi, Grasian, and Samuel Spiegel. 2020. "Mobility, Temporary Migration and Changing Livelihoods in Zimbabwe's Artisanal Mining Sector." *The Extractive Industries and Society* 7 (3): 994–1001. doi: 10.1016/j.exis.2020.05.001.

Owen, John R., and Deanna Kemp. 2017. "Social Management Capability, Human Migration and the Global Mining Industry." *Resources Policy* 53: 259–266. doi: 10.1016/j.resourpol.2017.06.017.

Prada-Trigo, Jose, Pablo Barra-Vieira, and Natalia Aravena-Solís. 2021. "Long-Distance Commuting and Real Estate Investment Linked to Mining: The Case Study of Concepción Metropolitan Area (Chile)." *Resources Policy* 70: 101973. doi: 10.1016/j.resourpol.2020.101973.

Rittel, Horst, and Melvin Webber. 1973. "Dilemmas in a General Theory of Planning." *Policy Sciences* 4: 155–169. doi: 10.1007/BF01405730.

Rodrigo, Luis, Gianni Romaní, and Emilio Ricci. 2018. "Immigrant Entrepreneurs in Antofagasta, Chile: The Contribution of South-South Entrepreneurs." *Academia, Revista Latinoamericana de Administración* 31 (3): 450–470. doi: 10.1108/ARLA-11-2016-0301.

Roumasset, James. 1976. *Rice and Risk: Decision Making Among Low-Income Farmers*. Amsterdam: North-Holland.

Schubert, Jon, Ulf Engel, and Elísio Macamo, eds. 2018. *Extractive Industries and Changing State Dynamics in Africa: Beyond the Resource Curse*. London: Routledge.

Scott, James C. 1976. *The Moral Economy of the Peasant: Rebellion and Subsistence in Southeast Asia*. New Haven: Yale University Press.

Wilson, Sigismond, Cyril Wilson, and Imelda Moise. 2022. "Livelihood Impacts of Iron Ore Mining-Induced Land Change in Sierra Leone: A Time Series Analysis." *Applied Geography* 144: 102713. doi: 10.1016/j.apgeog.2022.102713.

Index

Note: **Bold** page numbers refer to tables; *italic* page numbers refer to figures and page numbers followed by "n" denote endnotes.

For Product Safety Concerns and Information please contact our EU representative GPSR@taylorandfrancis.com Taylor & Francis Verlag GmbH, Kaufingerstraße 24, 80331 München, Germany

Printed and bound by CPI Group (UK) Ltd, Croydon, CR0 4YY
30/04/2025
01857439-0003